Use R!

Advisors:
Robert Gentleman ● Kurt Hornik ● Giovanni Parmigiani

For other titles published in this series, go to
http://www.springer.com/series/6991

Hadley Wickham

ggplot2

Elegant Graphics for Data Analysis

 Springer

Hadley Wickham
Rice University
Department of Statistics
Houston, TX
77005-1827
USA
hadley@rice.edu

Series Editors

Robert Gentleman
Program in Computational Biology
Division of Public Health Sciences
Fred Hutchinson Cancer Research Center
1100 Fairview Avenue, N. M2-B876
Seattle, Washington 98109
USA

Kurt Hornik
Department of Statistik and Mathematik
Wirtschaftsuniversität Wien Augasse 2-6
A-1090 Wien
Austria

Giovanni Parmigiani
The Sidney Kimmel Comprehensive Cancer
 Center at Johns Hopkins University
550 North Broadway
Baltimore, MD 21205-2011
USA

ISBN 978-0-387-98140-6   e-ISBN 978-0-387-98141-3
DOI 10.1007/978-0-387-98141-3
Springer Dordrecht Heidelberg London New York

Library of Congress Control Number: 2009928510

Printed on acid-free paper

Springer is part of Springer Science+Business Media (www.springer.com)

Contents

Appendices

Chapter 1

Introduction

1.1 Welcome to ggplot2

ggplot2 is an R package for producing statistical, or data, graphics, but
it is unlike most other graphics packages because it has a deep underlying
grammar. This grammar, based on the Grammar of Graphics (Wilkinson,
2005), is composed of a set of independent components that can be composed
in many different ways. This makes ggplot2 very powerful, because you are
not limited to a set of pre-specified graphics, but you can create new graphics
that are precisely tailored for your problem. This may sound overwhelming,
but because there is a simple set of core principles and very few special cases,
ggplot2 is also easy to learn (although it may take a little time to forget your
preconceptions from other graphics tools).

Practically, ggplot2 provides beautiful, hassle-free plots, that take care of
fiddly details like drawing legends. The plots can be built up iteratively and
edited later. A carefully chosen set of defaults means that most of the time
you can produce a publication-quality graphic in seconds, but if you do have
special formatting requirements, a comprehensive theming system makes it
easy to do what you want. Instead of spending time making your graph look
pretty, you can focus on creating a graph that best reveals the messages in
your data.

ggplot2 is designed to work in a layered fashion, starting with a layer
showing the raw data then adding layers of annotations and statistical summaries. It allows you to produce graphics using the same structured thinking
that you use to design an analysis, reducing the distance between a plot in
your head and one on the page. It is especially helpful for students who have
not yet developed the structured approach to analysis used by experts.

Learning the grammar will help you not only create graphics that you know
about now, but will also help you to think about new graphics that would be
even better. Without the grammar, there is no underlying theory and existing
graphics packages are just a big collection of special cases. For example, in
base R, if you design a new graphic, it's composed of raw plot elements like

H. Wickham, *ggplot2*, Use R, DOI 10.1007/978-0-387-98141-3_1,
© Springer Science+Business Media, LLC 2009

points and lines, and it's hard to design new components that combine with existing plots. In `ggplot2`, the expressions used to create a new graphic are composed of higher-level elements like representations of the raw data and statistical transformations, and can easily be combined with new datasets and other plots.

This book provides a hands-on introduction to `ggplot2` with lots of example code and graphics. It also explains the grammar on which `ggplot2` is based. Like other formal systems, `ggplot2` is useful even when you don't understand the underlying model. However, the more you learn about it, the more effectively you'll be able to use `ggplot2`. This book assumes some basic familiarity with R, to the level described in the first chapter of Dalgaard's *Introductory Statistics with R*. You should know how to get your data into R and how to do basic data manipulations. If you don't, you might want to get a copy of Phil Spector's *Data Manipulation with R*.

This book will introduce you to `ggplot2` as a novice, unfamiliar with the grammar; teach you the basics so that you can re-create plots you are already familiar with; show you how to use the grammar to create new types of graphics; and even turn you into an expert who can build new components to extend the grammar.

1.2 Other resources

This book teaches you the elements of `ggplot2`'s grammar and how they fit together, but it does not document every function in complete detail. Furthermore, `ggplot2` will almost certainly continue to evolve. For these reasons, you will need additional documentation as your use of `ggplot2` becomes more complex and varied.

The best resource for low-level details will always be the built-in documentation. This is accessible online, `http://had.co.nz/ggplot2`, and from within R using the usual help syntax. The advantage of the online documentation is that you can see all the example plots and navigate between topics more easily.

The website also lists talks and papers related to `ggplot2` and training opportunities if you'd like some hands-on practice. The CRAN website, `http://cran.r-project.org/web/packages/ggplot2/`, is another useful resource. This page links to what is new and different in each release. If you use `ggplot2` regularly, it's a good idea to sign up for the ggplot2 mailing list, `http://groups.google.com/group/ggplot2`. The list has relatively low traffic and is very friendly to new users.

Finally, the book website, `http://had.co.nz/ggplot2/book`, provides updates to this book, as well as pdfs containing all graphics used in the book, with the code and data needed to reproduce them.

1.3 What is the grammar of graphics?

Wilkinson (2005) created the grammar of graphics to describe the deep features that underlie all statistical graphics. The grammar of graphics is an answer to a question: what is a statistical graphic? The layered grammar of graphics (Wickham, 2009) builds on Wilkinson's grammar, focussing on the primacy of layers and adapting it for embedding within R. In brief, the grammar tells us that a statistical graphic is a mapping from data to aesthetic attributes (colour, shape, size) of geometric objects (points, lines, bars). The plot may also contain statistical transformations of the data and is drawn on a specific coordinate system. Faceting can be used to generate the same plot for different subsets of the dataset. It is the combination of these independent components that make up a graphic.

As the book progresses, the formal grammar will be explained in increasing detail. The first description of the components follows below. It introduces some of the terminology that will be used throughout the book and outlines the basic responsibilities of each component. Don't worry if it doesn't all make sense right away: you will have many more opportunities to learn about all of the pieces and how they fit together.

- The **data** that you want to visualise and a set of aesthetic **mapping**s describing how variables in the data are mapped to aesthetic attributes that you can perceive.
- Geometric objects, **geom**s for short, represent what you actually see on the plot: points, lines, polygons, etc.
- Statistical transformations, **stat**s for short, summarise data in many useful ways. For example, binning and counting observations to create a histogram, or summarising a 2d relationship with a linear model. Stats are optional, but very useful.
- The **scale**s map values in the data space to values in an aesthetic space, whether it be colour, or size, or shape. Scales draw a legend or axes, which provide an inverse mapping to make it possible to read the original data values from the graph.
- A coordinate system, **coord** for short, describes how data coordinates are mapped to the plane of the graphic. It also provides axes and gridlines to make it possible to read the graph. We normally use a Cartesian coordinate system, but a number of others are available, including polar coordinates and map projections.
- A **facet**ing specification describes how to break up the data into subsets and how to display those subsets as small multiples. This is also known as conditioning or latticing/trellising.

It is also important to talk about what the grammar doesn't do:

- It doesn't suggest what graphics you should use to answer the questions you are interested in. While this book endeavours to promote a sensible process for producing plots of data, the focus of the book is on how to produce the plots you want, not knowing what plots to produce. For more advice on this topic, you may want to consult Chambers et al. (1983); Cleveland (1993a); Robbins (2004); Tukey (1977).

- Ironically, the grammar doesn't specify what a graphic should look like. The finer points of display, for example, font size or background colour, are not specified by the grammar. In practice, a useful plotting system will need to describe these, as `ggplot2` does with its theming system. Similarly, the grammar does not specify how to make an attractive graphic and while the defaults in `ggplot2` have been chosen with care, you may need to consult other references to create an attractive plot: Tufte (1990, 1997, 2001, 2006).

- It does not describe interaction: the grammar of graphics describes only static graphics and there is essentially no benefit to displaying on a computer screen as opposed to on a piece of paper. `ggplot2` can only create static graphics, so for dynamic and interactive graphics you will have to look elsewhere. Cook and Swayne (2007) provides an excellent introduction to the interactive graphics package GGobi. GGobi can be connected to R with the `rggobi` package (Wickham et al., 2008).

1.4 How does `ggplot2` fit in with other R graphics?

There are a number of other graphics systems available in R: base graphics, grid graphics and trellis/lattice graphics. How does `ggplot2` differ from them?

- Base graphics were written by Ross Ihaka based on experience implementing S graphics driver and partly looking at Chambers et al. (1983). Base graphics has a pen on paper model: you can only draw on top of the plot, you cannot modify or delete existing content. There is no (user accessible) representation of the graphics, apart from their appearance on the screen. Base graphics includes both tools for drawing primitives and entire plots. Base graphics functions are generally fast, but have limited scope. When you've created a single scatterplot, or histogram, or a set of boxplots in the past, you've probably used base graphics.

- The development of `grid` graphics, a much richer system of graphical primitives, started in 2000. Grid is developed by Paul Murrell, growing out of his PhD work (Murrell, 1998). Grid grobs (graphical objects) can be represented independently of the plot and modified later. A system of viewports (each containing its own coordinate system) makes it easier to lay out complex graphics. Grid provides drawing primitives, but no tools for producing statistical graphics.

- The `lattice` package (Sarkar, 2008a), developed by Deepayan Sarkar, uses grid graphics to implement the trellis graphics system of Cleveland (1993a, 1985) and is a considerable improvement over base graphics. You can easily produce conditioned plots and some plotting details (e.g., legends) are taken care of automatically. However, lattice graphics lacks a formal model, which can make it hard to extend. Lattice graphics are explained in depth in (Sarkar, 2008b).

- `ggplot2`, started in 2005, is an attempt to take the good things about base and lattice graphics and improve on them with a strong underlying model which supports the production of any kind of statistical graphic, based on principles outlined above. The solid underlying model of `ggplot2` makes it easy to describe a wide range of graphics with a compact syntax and independent components make extension easy. Like `lattice`, `ggplot2` uses grid to draw the graphics, which means you can exercise much low-level control over the appearance of the plot.

Many other R packages, such as `vcd` (Meyer et al., 2006), `plotrix` (Lemon et al., 2008) and `gplots` (Warnes, 2007), implement specialist graphics, but no others provide a framework for producing statistical graphics. A comprehensive resource listing all graphics functionality available in other contributed packages is the graphics task view at `http://cran.r-project.org/web/views/Graphics.html`.

1.5 About this book

Chapter 2 describes how to quickly get started using `qplot` to make graphics, just like you can using `plot`. This chapter introduces several important `ggplot2` concepts: geoms, aesthetic mappings and faceting.

While `qplot` is a quick way to get started, you are not using the full power of the grammar. Chapter 3 describes the layered grammar of graphics which underlies `ggplot2`. The theory is illustrated in Chapter 4 which demonstrates how to add additional layers to your plot, exercising full control over the geoms and stats used within them. Chapter 5 describes how to assemble and combine geoms and stats to solve particular plotting problems.

Understanding how scales works is crucial for fine tuning the perceptual properties of your plot. Customising scales gives fine control over the exact appearance of the plot and helps to support the story that you are telling. Chapter 6 will show you what scales are available, how to adjust their parameters, and how to control the appearance of axes and legends.

Coordinate systems and faceting control the position of elements of the plot. These are described in Chapter 7. Faceting is a very powerful graphical tool as it allows you to rapidly compare different subsets of your data. Different coordinate systems are less commonly needed, but are very important for certain types of data.

To fine tune your plots for publication, you will need to learn about the tools described in Chapter 8. There you will learn about how to control the theming system of `ggplot2`, how to change the defaults for geoms, stats and scales, how to save plots to disk, and how to lay out multiple plots on a page.

The book concludes with two chapters that discuss high-level concerns about data structure and code duplication. Chapter 9 discusses some techniques that will enable you to get your data into the form required for `ggplot2`, and tools that enable you to perform more advanced aggregation and manipulation than is available in the plotting code. You will also learn about the `ggplot2` philosophy behind visualising other types of objects, and how you can extend `ggplot2` with your own methods.

Duplicated code is a big inhibitor of flexibility and reduces your ability to respond to changes in requirements. Chapter 10 covers three useful techniques for reducing duplication in your code: iteration, plot templates and plot functions.

Three appendices provide additional useful information. Appendix B describes how colours, shapes, line types and sizes can be specified by hand. Appendix A shows how to translate the syntax of base graphics, lattice graphics, and Wilkinson's GPL to `ggplot2` syntax. Appendix C describes the high-level organisation of grid objects and viewports used to draw a `ggplot2` plot. This will be useful if you are familiar with grid, and want to make changes to the underlying objects used to draw the plots.

1.6 Installation

To use `ggplot2`, you must first install it. Make sure you have a recent version of R (at least version 2.8) from `http://r-project.org` and then run the following line of code to download and install the `ggplot2` package.

```
install.packages("ggplot2")
```

`ggplot2` isn't perfect, so from time to time you may encounter something that doesn't work the way it should. If this happens, please email me at hadley@rice.edu with a reproducible example of your problem, as well as a description of what you think should have happened. The more information you provide, the easier it is for me to help you. .

1.7 Acknowledgements

Many people have contributed to this book with high-level structural insights, spelling and grammar corrections and bug reports. In particular, I would like to thank: Leland Wilkinson, for discussions and comments that cemented my understanding of the grammar; Gabor Grothendieck, for early helpful comments; Heike Hofmann and Di Cook, for being great major professors; Charlotte Wickham; the students of stat480 and stat503 at ISU, for trying it out when it was very young; Debby Swayne, for masses of helpful feedback and advice; Bob Muenchen, Reinhold Kliegl, Philipp Pagel, Richard Stahlhut,

Baptiste Auguie, Jean-Olivier Irisson, Thierry Onkelinx and the many others who have read draft versions of the book and given me feedback; and last, but not least, the members of R-help and the `ggplot2` mailing list, for providing the many interesting and challenging graphics problems that have helped motivate this book.

Chapter 2

Getting started with qplot

2.1 Introduction

In this chapter, you will learn to make a wide variety of plots with your first ggplot2 function, qplot(), short for quick plot. qplot makes it easy to produce complex plots, often requiring several lines of code using other plotting systems, in one line. qplot() can do this because it's based on the grammar of graphics, which allows you to create a simple, yet expressive, description of the plot. In later chapters you'll learn to use all of the expressive power of the grammar, but here we'll start simple so you can work your way up. You will also start to learn some of the ggplot2 terminology that will be used throughout the book.

qplot has been designed to be very similar to plot, which should make it easy if you're already familiar with plotting in R. Remember, during an R session you can get a summary of all the arguments to qplot with R help, ?qplot.

In this chapter you'll learn:

- The basic use of qplot—If you're already familiar with plot, this will be particularly easy, § 2.3.
- How to map variables to aesthetic attributes, like colour, size and shape, § 2.4.
- How to create many different types of plots by specifying different geoms, and how to combine multiple types in a single plot, § 2.5.
- The use of faceting, also known as trellising or conditioning, to break apart subsets of your data, § 2.6.
- How to tune the appearance of the plot by specifying some basic options, § 2.7.
- A few important differences between plot() and qplot(), § 2.8.

H. Wickham, *ggplot2*, Use R, DOI 10.1007/978-0-387-98141-3_2,
© Springer Science+Business Media, LLC 2009

2.2 Datasets

In this chapter we'll just use one data source, so you can get familiar with the plotting details rather than having to familiarise yourself with different datasets. The `diamonds` dataset consists of prices and quality information about 54,000 diamonds, and is included in the **ggplot2** package. The data contains the four C's of diamond quality, carat, cut, colour and clarity; and five physical measurements, depth, table, x, y and z, as described in Figure 2.1. The first few rows of the data are shown in Table 2.1.

carat	cut	color	clarity	depth	table	price	x	y	z
0.2	Ideal	E	SI2	61.5	55.0	326	3.95	3.98	2.43
0.2	Premium	E	SI1	59.8	61.0	326	3.89	3.84	2.31
0.2	Good	E	VS1	56.9	65.0	327	4.05	4.07	2.31
0.3	Premium	I	VS2	62.4	58.0	334	4.20	4.23	2.63
0.3	Good	J	SI2	63.3	58.0	335	4.34	4.35	2.75
0.2	Very Good	J	VVS2	62.8	57.0	336	3.94	3.96	2.48

Table 2.1: `diamonds` dataset. The variables depth, table, x, y and z refer to the dimensions of the diamond as shown in Figure 2.1

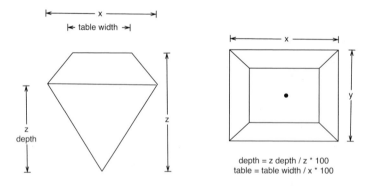

Fig. 2.1: How the variables x, y, z, table and depth are measured.

The dataset has not been well cleaned, so as well as demonstrating interesting relationships about diamonds, it also demonstrates some data quality problems. We'll also use another dataset, `dsmall`, which is a random sample of 100 diamonds. We'll use this data for plots that are more appropriate for smaller datasets.

```
> set.seed(1410) # Make the sample reproducible
```

```
> dsmall <- diamonds[sample(nrow(diamonds), 100), ]
```

2.3 Basic use

As with **plot**, the first two arguments to **qplot()** are x and y, giving the
x- and y-coordinates for the objects on the plot. There is also an optional
data argument. If this is specified, **qplot()** will look inside that data frame
before looking for objects in your workspace. Using the **data** argument is
recommended: it's a good idea to keep related data in a single data frame. If
you don't specify one, **qplot()** will try to build one up for you and may look
in the wrong place.

Here is a simple example of the use of **qplot()**. It produces a scatterplot
showing the relationship between the price and carats (weight) of a diamond.

```
> qplot(carat, price, data = diamonds)
```

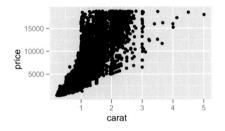

The plot shows a strong correlation with notable outliers and some interesting vertical striation. The relationship looks exponential, though, so the first
thing we'd like to do is to transform the variables. Because **qplot()** accepts
functions of variables as arguments, we plot log(price) vs. log(carat):

```
> qplot(log(carat), log(price), data = diamonds)
```

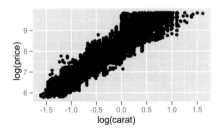

The relationship now looks linear. With this much overplotting, though, we
need to be cautious about drawing firm conclusions.

Arguments can also be combinations of existing variables, so, if we are curi-
ous about the relationship between the volume of the diamond (approximated
by $x \times y \times z$) and its weight, we could do the following:

```
> qplot(carat, x * y * z, data = diamonds)
```

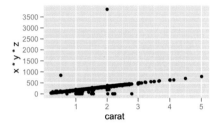

We would expect the density (weight/volume) of diamonds to be constant, and so see a linear relationship between volume and weight. The majority of diamonds do seem to fall along a line, but there are some large outliers.

2.4 Colour, size, shape and other aesthetic attributes

The first big difference when using qplot instead of plot comes when you want to assign colours—or sizes or shapes—to the points on your plot. With plot, it's your responsibility to convert a categorical variable in your data (e.g., "apples", "bananas", "pears") into something that plot knows how to use (e.g., "red", "yellow", "green"). qplot can do this for you automatically, and it will automatically provide a legend that maps the displayed attributes to the data values. This makes it easy to include additional data on the plot.

In the next example, we augment the plot of carat and price with information about diamond colour and cut. The results are shown in Figure 2.2.

```
qplot(carat, price, data = dsmall, colour = color)
qplot(carat, price, data = dsmall, shape = cut)
```

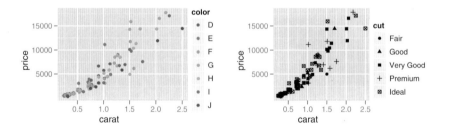

Fig. 2.2: Mapping point colour to diamond colour (left), and point shape to cut quality (right).

Colour, size and shape are all examples of aesthetic attributes, visual properties that affect the way observations are displayed. For every aesthetic attribute, there is a function, called a *scale*, which maps data values to valid values for that aesthetic. It is this scale that controls the appearance of the points and associated legend. For example, in the above plots, the colour scale maps J to purple and F to green. (Note that while I use British spelling throughout this book, the software also accepts American spellings.)

You can also manually set the aesthetics using `I()`, e.g., `colour = I("red")` or `size = I(2)`. This is not the same as mapping and is explained in more detail in Section 4.5.2. For large datasets, like the diamonds data, semi-transparent points are often useful to alleviate some of the overplotting. To make a semi-transparent colour you can use the alpha aesthetic, which takes a value between 0 (completely transparent) and 1 (complete opaque). It's often useful to specify the transparency as a fraction, e.g., 1/10 or 1/20, as the denominator specifies the number of points that must overplot to get a completely opaque colour.

```
qplot(carat, price, data = diamonds, alpha = I(1/10))
qplot(carat, price, data = diamonds, alpha = I(1/100))
qplot(carat, price, data = diamonds, alpha = I(1/200))
```

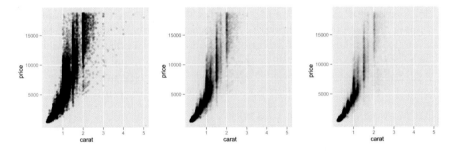

Fig. 2.3: Reducing the alpha value from 1/10 (left) to 1/100 (middle) to 1/200 (right) makes it possible to see where the bulk of the points lie.

Different types of aesthetic attributes work better with different types of variables. For example, colour and shape work well with categorical variables, while size works better with continuous variables. The amount of data also makes a difference: if there is a lot of data, like in the plots above, it can be hard to distinguish the different groups. An alternative solution is to use faceting, which will be introduced in Section 2.6.

2.5 Plot geoms

`qplot` is not limited to scatterplots, but can produce almost any kind of plot by varying the `geom`. Geom, short for geometric object, describes the type of object that is used to display the data. Some geoms have an associated statistical transformation, for example, a histogram is a binning statistic plus a bar geom. These different components are described in the next chapter. Here we'll introduce the most common and useful geoms, organised by the dimensionality of data that they work with. The following geoms enable you to investigate two-dimensional relationships:

- `geom` = `"point"` draws points to produce a scatterplot. This is the default when you supply both `x` and `y` arguments to `qplot()`.
- `geom` = `"smooth"` fits a smoother to the data and displays the smooth and its standard error, § 2.5.1.
- `geom` = `"boxplot"` produces a box-and-whisker plot to summarise the distribution of a set of points, § 2.5.2.
- `geom` = `"path"` and `geom` = `"line"` draw lines between the data points. Traditionally these are used to explore relationships between time and another variable, but lines may be used to join observations connected in some other way. A line plot is constrained to produce lines that travel from left to right, while paths can go in any direction, § 2.5.5.

For 1d distributions, your choice of geoms is guided by the variable type:

- For continuous variables, `geom` = `"histogram"` draws a histogram, `geom` = `"freqpoly"` a frequency polygon, and `geom` = `"density"` creates a density plot, § 2.5.3. The histogram geom is the default when you only supply an `x` value to `qplot()`.
- For discrete variables, `geom` = `"bar"` makes a bar chart, § 2.5.4.

2.5.1 Adding a smoother to a plot

If you have a scatterplot with many data points, it can be hard to see exactly what trend is shown by the data. In this case you may want to add a smoothed line to the plot. This is easily done using the `smooth` geom as shown in Figure 2.4. Notice that we have combined multiple geoms by supplying a vector of geom names created with `c()`. The geoms will be overlaid in the order in which they appear.

```
qplot(carat, price, data = dsmall, geom = c("point", "smooth"))
qplot(carat, price, data = diamonds, geom = c("point", "smooth"))
```

Despite overplotting, our impression of an exponential relationship between price and carat was correct. There are few diamonds bigger than three carats, and our uncertainty in the form of the relationship increases as illustrated by the point-wise confidence interval shown in grey. If you want to turn the confidence interval off, use `se` = `FALSE`.

There are many different smoothers you can choose between by using the `method` argument:

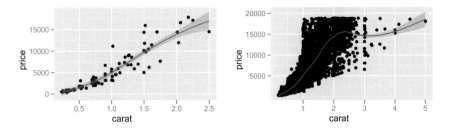

Fig. 2.4: Smooth curves add to scatterplots of carat vs. price. The dsmall dataset (left) and the full dataset (right).

- `method = "loess"`, the default for small n, uses a smooth local regression. More details about the algorithm used can be found in `?loess`. The wiggliness of the line is controlled by the `span` parameter, which ranges from 0 (exceedingly wiggly) to 1 (not so wiggly), as shown in Figure 2.5.

```
qplot(carat, price, data = dsmall, geom = c("point", "smooth"),
  span = 0.2)
qplot(carat, price, data = dsmall, geom = c("point", "smooth"),
  span = 1)
```

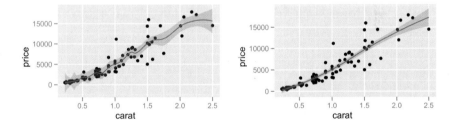

Fig. 2.5: The effect of the span parameter. (Left) `span = 0.2`, and (right) `span = 1`.

Loess does not work well for large datasets (it's $O(n^2)$ in memory), and so an alternative smoothing algorithm is used when n is greater than 1,000.

- You could also load the `mgcv` library and use `method = "gam"`, `formula = y ~ s(x)` to fit a generalised additive model. This is similar to using a spline with `lm`, but the degree of smoothness is estimated from the data. For large data, use the formula `y ~ s(x, bs = "cs")`. This is used by default when there are more than 1,000 points.

```
library(mgcv)
qplot(carat, price, data = dsmall, geom = c("point", "smooth"),
```

```
    method = "gam", formula = y ~ s(x))
  qplot(carat, price, data = dsmall, geom = c("point", "smooth"),
    method = "gam", formula = y ~ s(x, bs = "cs"))
```

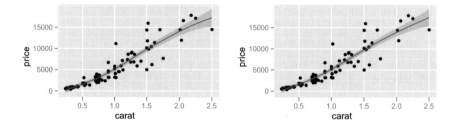

Fig. 2.6: The effect of the formula parameter, using a generalised additive model as a smoother. (Left) formula = y ~ s(x), the default; (right) formula = y ~ s(x, bs = "cs").

- method = "lm" fits a linear model. The default will fit a straight line to your data, or you can specify formula = y ~ poly(x, 2) to specify a degree 2 polynomial, or better, load the splines package and use a natural spline: formula = y ~ ns(x, 2). The second parameter is the degrees of freedom: a higher number will create a wigglier curve. You are free to specify any formula involving x and y. Figure 2.7 shows two examples created with the following code.

```
library(splines)
qplot(carat, price, data = dsmall, geom = c("point", "smooth"),
  method = "lm")
qplot(carat, price, data = dsmall, geom = c("point", "smooth"),
  method = "lm", formula = y ~ ns(x,5))
```

- method = "rlm" works like lm, but uses a robust fitting algorithm so that outliers don't affect the fit as much. It's part of the MASS package, so remember to load that first.

2.5.2 Boxplots and jittered points

When a set of data includes a categorical variable and one or more continuous variables, you will probably be interested to know how the values of the continuous variables vary with the levels of the categorical variable. Boxplots and jittered points offer two ways to do this. Figure 2.8 explores how the distribution of price per carat varies with the colour of the diamond using jittering (geom = "jitter", left) and box-and-whisker plots (geom = "boxplot", right).

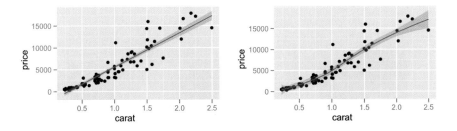

Fig. 2.7: The effect of the formula parameter, using a linear model as a smoother. (Left) `formula = y ~ x`, the default; (right) `formula = y ~ ns(x, 5)`.

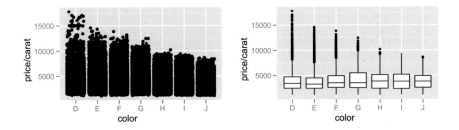

Fig. 2.8: Using jittering (left) and boxplots (right) to investigate the distribution of price per carat, conditional on colour. As the colour improves (from left to right) the spread of values decreases, but there is little change in the centre of the distribution.

Each method has its strengths and weaknesses. Boxplots summarise the bulk of the distribution with only five numbers, while jittered plots show every point but can suffer from overplotting. In the example here, both plots show the dependency of the spread of price per carat on diamond colour, but the boxplots are more informative, indicating that there is very little change in the median and adjacent quartiles.

The overplotting seen in the plot of jittered values can be alleviated somewhat by using semi-transparent points using the `alpha` argument. Figure 2.9 illustrates three different levels of transparency, which make it easier to see where the bulk of the points lie. The plots are produced with the following code.

```
qplot(color, price / carat, data = diamonds, geom = "jitter",
  alpha = I(1 / 5))
qplot(color, price / carat, data = diamonds, geom = "jitter",
  alpha = I(1 / 50))
qplot(color, price / carat, data = diamonds, geom = "jitter",
  alpha = I(1 / 200))
```

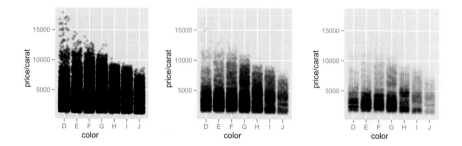

Fig. 2.9: Varying the alpha level. From left to right: 1/5, 1/50, 1/200. As the opacity decreases we begin to see where the bulk of the data lies. However, the boxplot still does much better.

This technique can't show the positions of the quantiles as well as a boxplot can, but it may reveal other features of the distribution that a boxplot cannot.

For jittered points, `qplot` offers the same control over aesthetics as it does for a normal scatterplot: `size`, `colour` and `shape`. For boxplots you can control the outline `colour`, the internal `fill` colour and the `size` of the lines.

Another way to look at conditional distributions is to use faceting to plot a separate histogram or density plot for each value of the categorical variable. This is demonstrated in Section 2.6.

2.5.3 Histogram and density plots

Histogram and density plots show the distribution of a single variable. They provide more information about the distribution of a single group than boxplots do, but it is harder to compare many groups (although we will look at one way to do so). Figure 2.10 shows the distribution of carats with a histogram and a density plot.

```
qplot(carat, data = diamonds, geom = "histogram")
qplot(carat, data = diamonds, geom = "density")
```

For the density plot, the `adjust` argument controls the degree of smoothness (high values of `adjust` produce smoother plots). For the histogram, the `binwidth` argument controls the amount of smoothing by setting the bin size. (Break points can also be specified explicitly, using the `breaks` argument.) It is **very important** to experiment with the level of smoothing. With a histogram you should try many bin widths: You may find that gross features of the data show up well at a large bin width, while finer features require a very narrow width.

In Figure 2.11, we experiment with three values of `binwidth`: 1.0, 0.1 and 0.01. It is only in the plot with the smallest bin width (right) that we see the striations we noted in an earlier scatterplot, most at "nice" numbers of carats. The full code is:

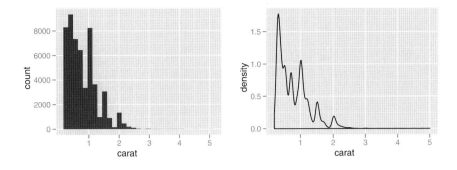

Fig. 2.10: Displaying the distribution of diamonds. (Left) geom = "histogram" and (right) geom = "density".

```
qplot(carat, data = diamonds, geom = "histogram", binwidth = 1,
   xlim = c(0,3))
qplot(carat, data = diamonds, geom = "histogram", binwidth = 0.1,
   xlim = c(0,3))
qplot(carat, data = diamonds, geom = "histogram", binwidth = 0.01,
   xlim = c(0,3))
```

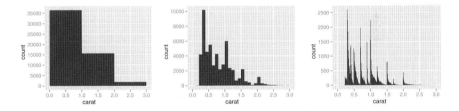

Fig. 2.11: Varying the bin width on a histogram of carat reveals interesting patterns. Binwidths from left to right: 1, 0.1 and 0.01 carats. Only diamonds between 0 and 3 carats shown.

To compare the distributions of different subgroups, just add an aesthetic mapping, as in the following code.

```
qplot(carat, data = diamonds, geom = "density", colour = color)
qplot(carat, data = diamonds, geom = "histogram", fill = color)
```

Mapping a categorical variable to an aesthetic will automatically split up the geom by that variable, so these commands instruct qplot() to draw a density plot and histogram for each level of diamond colour. The results are shown in Figure 2.12.

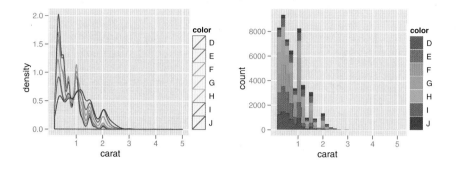

Fig. 2.12: Mapping a categorical variable to an aesthetic will automatically split up the geom by that variable. (Left) Density plots are overlaid and (right) histograms are stacked.

The density plot is more appealing at first because it seems easy to read and compare the various curves. However, it is more difficult to understand exactly what a density plot is showing. In addition, the density plot makes some assumptions that may not be true for our data; i.e., that it is unbounded, continuous and smooth.

2.5.4 Bar charts

The discrete analogue of histogram is the bar chart, `geom = "bar"`. The bar geom counts the number of instances of each class so that you don't need to tabulate your values beforehand, as with `barchart` in base R. If the data has already been tabulated or if you'd like to tabulate class members in some other way, such as by summing up a continuous variable, you can use the `weight` geom. This is illustrated in Figure 2.13. The first plot is a simple bar chart of diamond colour, and the second is a bar chart of diamond colour weighted by carat.

```
qplot(color, data = diamonds, geom = "bar")
qplot(color, data = diamonds, geom = "bar", weight = carat) +
  scale_y_continuous("carat")
```

2.5.5 Time series with line and path plots

Line and path plots are typically used for time series data. Line plots join the points from left to right, while path plots join them in the order that they appear in the dataset (a line plot is just a path plot of the data sorted by x value). Line plots usually have time on the x-axis, showing how a single variable has changed over time. Path plots show how two variables have simultaneously

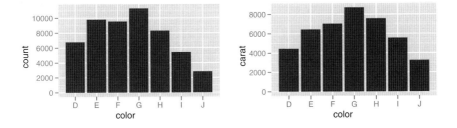

Fig. 2.13: Bar charts of diamond colour. The left plot shows counts and the right plot is weighted by `weight = carat` to show the total weight of diamonds of each colour.

changed over time, with time encoded in the way that the points are joined together.

Because there is no time variable in the diamonds data, we use the `economics` dataset, which contains economic data on the US measured over the last 40 years. Figure 2.14 shows two plots of unemployment over time, both produced using `geom = "line"`. The first shows an unemployment rate and the second shows the median number of weeks unemployed. We can already see some differences in these two variables, particularly in the last peak, where the unemployment percentage is lower than it was in the preceding peaks, but the length of unemployment is high.

```
qplot(date, unemploy / pop, data = economics, geom = "line")
qplot(date, uempmed, data = economics, geom = "line")
```

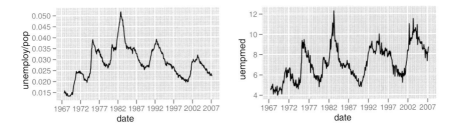

Fig. 2.14: Two time series measuring amount of unemployment. (Left) Percent of population that is unemployed and (right) median number of weeks unemployed. Plots created with `geom="line"`.

To examine this relationship in greater detail, we would like to draw both time series on the same plot. We could draw a scatterplot of unemployment rate vs. length of unemployment, but then we could no longer see the evolution over time. The solution is to join points adjacent in time with line segments, forming a *path* plot.

Below we plot unemployment rate vs. length of unemployment and join the individual observations with a path. Because of the many line crossings, the direction in which time flows isn't easy to see in the first plot. In the second plot, we apply the `colour` aesthetic to the line to make it easier to see the direction of time.

```
year <- function(x) as.POSIXlt(x)$year + 1900
qplot(unemploy / pop, uempmed, data = economics,
   geom = c("point", "path"))
qplot(unemploy / pop, uempmed, data = economics,
  geom = "path", colour = year(date)) + scale_area()
```

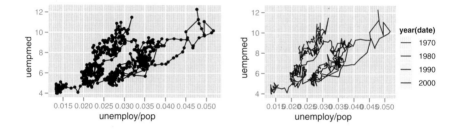

Fig. 2.15: Path plots illustrating the relationship between percent of people unemployed and median length of unemployment. (Left) Scatterplot with overlaid path. (Right) Pure path plot coloured by year.

We can see that percent unemployed and length of unemployment are highly correlated, although in recent years the length of unemployment has been increasing relative to the unemployment rate.

With longitudinal data, you often want to display multiple time series on each plot, each series representing one individual. To do this with `qplot()`, you need to map the `group` aesthetic to a variable encoding the group membership of each observation. This is explained in more depth in Section 4.5.3.

2.6 Faceting

We have already discussed using aesthetics (colour and shape) to compare subgroups, drawing all groups on the same plot. Faceting takes an alternative approach: It creates tables of graphics by splitting the data into subsets and displaying the same graph for each subset in an arrangement that facilitates comparison. Section 7.2 discusses faceting in detail, including a discussion of the advantages and disadvantages of using faceting instead of aesthetics in Section 7.2.5.

The default faceting method in `qplot()` creates plots arranged on a grid specified by a faceting formula which looks like `row_var ~ col_var`. You can specify as many row and column variables as you like, keeping in mind that using more than two variables will often produce a plot so large that it is difficult to see on screen. To facet on only one of columns or rows, use . as a place holder. For example, `row_var ~ .` will create a single column with multiple rows.

Figure 2.16 illustrates this technique with two plots, sets of histograms showing the distribution of carat conditional on colour. The second set of histograms shows proportions, making it easier to compare distributions regardless of the relative abundance of diamonds of each colour. The `..density..` syntax is new. The y-axis of the histogram does not come from the original data, but from the statistical transformation that counts the number of observations in each bin. Using `..density..` tells ggplot2 to map the density to the y-axis instead of the default use of count.

```
qplot(carat, data = diamonds, facets = color ~ .,
  geom = "histogram", binwidth = 0.1, xlim = c(0, 3))
qplot(carat, ..density.., data = diamonds, facets = color ~ .,
  geom = "histogram", binwidth = 0.1, xlim = c(0, 3))
```

2.7 Other options

These are a few other `qplot` options to control the graphic's appearance. These all have the same effect as their `plot` equivalents:

- `xlim`, `ylim`: set limits for the x- and y-axes, each a numeric vector of length two, e.g., `xlim=c(0, 20)` or `ylim=c(-0.9, -0.5)`.
- `log`: a character vector indicating which (if any) axes should be logged. For example, `log="x"` will log the x-axis, `log="xy"` will log both.
- `main`: main title for the plot, centered in large text at the top of the plot. This can be a string (e.g., `main="plot title"`) or an expression (e.g., `main = expression(beta[1] == 1)`). See `?plotmath` for more examples of using mathematical formulae.
- `xlab`, `ylab`: labels for the x- and y-axes. As with the plot title, these can be character strings or mathematical expressions.

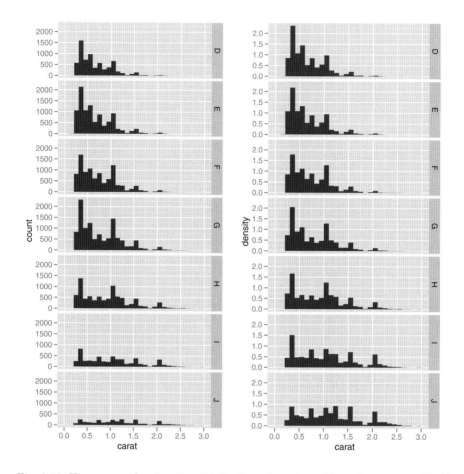

Fig. 2.16: Histograms showing the distribution of carat conditional on colour. (Left) Bars show counts and (right) bars show densities (proportions of the whole). The density plot makes it easier to compare distributions ignoring the relative abundance of diamonds within each colour. High-quality diamonds (colour D) are skewed towards small sizes, and as quality declines the distribution becomes more flat.

The following examples show the options in action.

```
> qplot(
+    carat, price, data = dsmall,
+    xlab = "Price ($)", ylab = "Weight (carats)",
+    main = "Price-weight relationship"
+ )
```

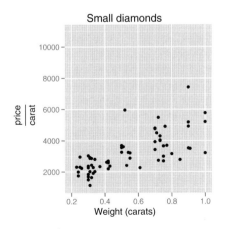

```
> qplot(
+    carat, price/carat, data = dsmall,
+    ylab = expression(frac(price,carat)),
+    xlab = "Weight (carats)",
+    main="Small diamonds",
+    xlim = c(.2,1)
+ )
WARNING: Removed 35 rows containing missing values (geom_point).
```

```
> qplot(carat, price, data = dsmall, log = "xy")
```

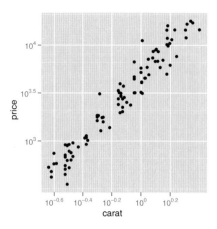

2.8 Differences from plot

There are a few important differences between `plot` and `qplot`:

- `qplot` is not generic: you cannot pass any type of R object to qplot and expect to get some kind of default plot. Note, however, that `ggplot()` is generic, and may provide a starting point for producing visualisations of arbitrary R objects. See Chapter 9 for more details.
- Usually you will supply a variable to the aesthetic attribute you're interested in. This is then scaled and displayed with a legend. If you want to set the value, e.g., to make red points, use `I()`: `colour = I("red")`. This is explained in more detail in Section 4.5.2.
- While you can continue to use the base R aesthetic names (`col`, `pch`, `cex`, etc.), it's a good idea to switch to the more descriptive `ggplot2` aesthetic names (`colour`, `shape` and `size`). They're much easier to remember!
- To add further graphic elements to a plot produced in base graphics, you can use `points()`, `lines()` and `text()`. With `ggplot2`, you need to add additional **layers** to the existing plot, described in the next chapter.

Chapter 3

Mastering the grammar

3.1 Introduction

You can choose to use just `qplot()`, without any understanding of the under-lying grammar, but if you do you will never be able to unlock the full power of `ggplot2`. By learning more about the grammar and its components, you will be able to create a wider range of plots, as well as being able to combine multiple sources of data, and customise to your heart's content. You may want to skip this chapter in a first reading of the book, returning when you want a deeper understanding of how all the pieces fit together.

This chapter describes the theoretical basis of `ggplot2`: the layered gram-mar of graphics. The layered grammar is based on Wilkinson's grammar of graphics (Wilkinson, 2005), but adds a number of enhancements that help it to be more expressive and fit seamlessly into the R environment. The differences between the layered grammar and Wilkinson's grammar are described fully in (Wickham, 2008), and a guide for converting between GPL (the encoding of the grammar used in SPSS) and `ggplot2` is included in Appendix A. In this chapter you will learn a little bit about each component of the grammar and how they all fit together. The next chapters discuss the components in more detail, and provide more examples of how you can use them in practice.

The grammar is useful for you both as a user and as a potential developer of statistical graphics. As a user, it makes it easier for you to iteratively update a plot, changing a single feature at a time. The grammar is also useful because it suggests the high-level aspects of a plot that *can* be changed, giving you a framework to think about graphics, and hopefully shortening the distance from mind to paper. It also encourages the use of graphics customised to a particular problem, rather than relying on generic named graphics.

As a developer, the grammar makes it much easier to add new capabilities to `ggplot2`. You only need to add the one component that you need, and you can continue to use all of the other existing components. For example, you can add a new statistical transformation, and continue to use the existing scales and geoms. It is also useful for discovering new types of graphics, as the grammar effectively defines the parameter space of statistical graphics.

H. Wickham, *ggplot2*, Use R, DOI 10.1007/978-0-387-98141-3_3,
© Springer Science+Business Media, LLC 2009

This chapter begins by describing in detail the process of drawing a simple plot. Section 3.3 starts with a simple scatterplot, then Section 3.4 makes it more complex by adding a smooth line and faceting. While working through these examples you will be introduced to all six components of the grammar, which are then defined more precisely in Section 3.5. The chapter concludes with Section 3.6, which describes how the various components map to data structures in R.

3.2 Fuel economy data

Consider the fuel economy dataset, mpg, a sample of which is illustrated in Table 3.1. It records make, model, class, engine size, transmission and fuel economy for a selection of US cars in 1999 and 2008. It contains the 38 models that were updated every year, an indicator that the car was a popular model. These models include popular cars like the Audi A4, Honda Civic, Hyundai Sonata, Nissan Maxima, Toyota Camry and Volkswagen Jetta. This data comes from the EPA fuel economy website, http://fueleconomy.gov.

manufacturer	model	disp	year	cyl	cty	hwy	class
audi	a4	1.8	1999	4	18	29	compact
audi	a4	1.8	1999	4	21	29	compact
audi	a4	2.0	2008	4	20	31	compact
audi	a4	2.0	2008	4	21	30	compact
audi	a4	2.8	1999	6	16	26	compact
audi	a4	2.8	1999	6	18	26	compact
audi	a4	3.1	2008	6	18	27	compact
audi	a4 quattro	1.8	1999	4	18	26	compact
audi	a4 quattro	1.8	1999	4	16	25	compact
audi	a4 quattro	2.0	2008	4	20	28	compact

Table 3.1: The first 10 cars in the mpg dataset, included in the ggplot2 package. cty and hwy record miles per gallon (mpg) for city and highway driving, respectively, and displ is the engine displacement in litres.

This dataset suggests many interesting questions. How are engine size and fuel economy related? Do certain manufacturers care more about economy than others? Has fuel economy improved in the last ten years? We will try to answer the first question and in the process learn more details about how the scatterplot is created.

3.3 Building a scatterplot

Consider Figure 3.1, one attempt to answer this question. It is a scatterplot of two continuous variables (engine displacement and highway mpg), with points coloured by a third variable (number of cylinders). From your experience in the previous chapter, you should have a pretty good feel for how to create this plot with `qplot()`. But what is going on underneath the surface? How does ggplot2 draw this plot?

```
qplot(displ, hwy, data = mpg, colour = factor(cyl))
```

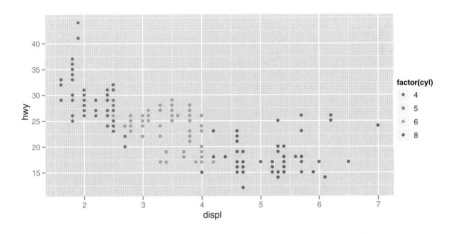

Fig. 3.1: A scatterplot of engine displacement in litres (displ) vs. average highway miles per gallon (hwy). Points are coloured according to number of cylinders. This plot summarises the most important factor governing fuel economy: engine size.

Mapping aesthetics to data

What precisely is a scatterplot? You have seen many before and have probably even drawn some by hand. A scatterplot represents each observation as a point (●), positioned according to the value of two variables. As well as a horizontal and vertical position, each point also has a size, a colour and a shape. These attributes are called **aesthetics**, and are the properties that can be perceived on the graphic. Each aesthetic can be mapped to a variable, or set to a constant value. In Figure 3.1 displ is mapped to horizontal position, hwy to vertical position and cyl to colour. Size and shape are not mapped to variables, but remain at their (constant) default values.

Once we have these mappings we can create a new dataset that records this information. Table 3.2 shows the first 10 rows of the data behind Figure 3.1.

This new dataset is a result of applying the aesthetic mappings to the original data. We can create many different types of plots using this data. The scatterplot uses points, but were we instead to draw lines we would get a line plot. If we used bars, we'd get a bar plot. Neither of those examples makes sense for this data, but we could still draw them, as in Figure 3.2. In ggplot2 we can produce many plots that don't make sense, yet are grammatically valid. This is no different than English, where we can create senseless but grammatical sentences like the angry rock barked like a comma.

x	y	colour
1.8	29	4
1.8	29	4
2.0	31	4
2.0	30	4
2.8	26	6
2.8	26	6
3.1	27	6
1.8	26	4
1.8	25	4
2.0	28	4

Table 3.2: First 10 rows from mpg rearranged into the format required for a scatterplot. This data frame contains all the data to be displayed on the plot.

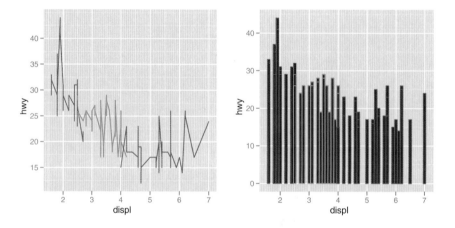

Fig. 3.2: Instead of using points to represent the data, we could use other geoms like lines (left) or bars (right). Neither of these geoms makes sense for this data, but they are still grammatically valid.

Points, lines and bars are all examples of geometric objects, or **geom**s. Geoms determine the "type" of the plot. Plots that use a single geom are often given a special name, a few of which are listed in Table 3.3. More complex plots with combinations of multiple geoms don't have a special name, and we have to describe them by hand. For example, Figure 3.3 overlays a per group regression line on the existing plot. What would you call this plot? Once you've mastered the grammar, you'll find that many of the plots that you produce are uniquely tailored to your problems and will no longer have special names.

Named plot	Geom	Other features
scatterplot	point	
bubblechart	point	size mapped to a variable
barchart	bar	
box-and-whisker plot	boxplot	
line chart	line	

Table 3.3: A selection of named plots and the geoms that they correspond to.

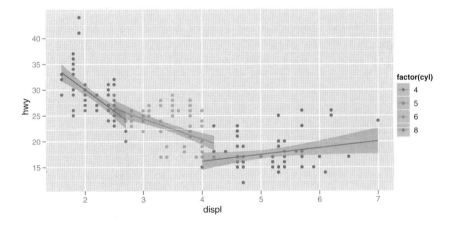

Fig. 3.3: More complicated plots don't have their own names. This plot takes Figure 3.1 and adds a regression line to each group. What would you call this plot?

Scaling

The values in Table 3.2 have no meaning to the computer. We need to convert them from data units (e.g., litres, miles per gallon and number of cylinders) to physical units (e.g., pixels and colours) that the computer can display. This conversion process is called **scaling** and performed by scales. Now that these values are meaningful to the computer, they may not be meaningful to us: colours are represented by a six-letter hexadecimal string, sizes by a number and shapes by an integer. These aesthetic specifications that are meaningful to R are described in Appendix B.

In this example, we have three aesthetics that need to be scaled: horizontal position (x), vertical position (y) and colour. Scaling position is easy in this example because we are using the default linear scales. We need only a linear mapping from the range of the data to $[0, 1]$. We use $[0, 1]$ instead of exact pixels because the drawing system that `ggplot2` uses, `grid`, takes care of that final conversion for us. A final step determines how the two positions (x and y) are combined to form the final location on the plot. This is done by the coordinate system, or **coord**. In most cases this will be Cartesian coordinates, but it might be polar coordinates, or a spherical projection used for a map.

The process for mapping the colour is a little more complicated, as we have a non-numeric result: colours. However, colours can be thought of as having three components, corresponding to the three types of colour-detecting cells in the human eye. These three cell types give rise to a three-dimensional colour space. Scaling then involves mapping the data values to points in this space. There are many ways to do this, but here since cyl is a categorical variable we map values to evenly spaced hues on the colour wheel, as shown in Figure 3.4. A different mapping is used when the variable is continuous.

The result of these conversions is Table 3.4, which contains values that have meaning to the computer. As well as aesthetics that have been mapped to variable, we also include aesthetics that are constant. We need these so that the aesthetics for each point are completely specified and R can draw the plot.

x	y	colour	size	shape
0.037	0.531	#FF6C91	1	19
0.037	0.531	#FF6C91	1	19
0.074	0.594	#FF6C91	1	19
0.074	0.562	#FF6C91	1	19
0.222	0.438	#00C1A9	1	19
0.222	0.438	#00C1A9	1	19
0.278	0.469	#00C1A9	1	19
0.037	0.438	#FF6C91	1	19
0.037	0.406	#FF6C91	1	19
0.074	0.500	#FF6C91	1	19

Table 3.4: Simple dataset with variables mapped into aesthetic space. The description of colours is intimidating, but this is the form that R uses internally. Default values for other aesthetics are filled in: the points will be filled circles (shape 19 in R) with a 1-mm diameter.

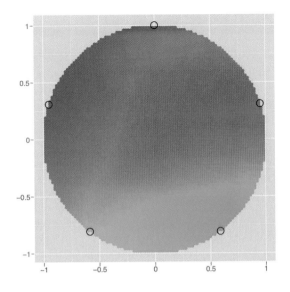

Fig. 3.4: A colour wheel illustrating the choice of five equally spaced colours. This is the default scale for discrete variables.

Finally, we need to render this data to create the graphical objects that are displayed on the screen. To create a complete plot we need to combine graphical objects from three sources: the *data*, represented by the point geom; the *scales and coordinate system*, which generate axes and legends so that we can read values from the graph; and *plot annotations*, such as the background and plot title. Figure 3.5 separates the contribution of the data from the contributions of the scales and plot annotations.

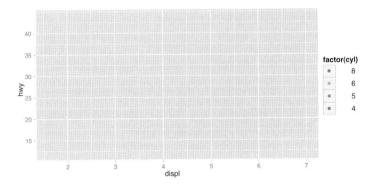

Fig. 3.5: Contributions from the scales, the axes and legend and grid lines, and the plot background. Contributions from the data, the point geom, have been removed.

3.4 A more complex plot

With a simple example under our belts, let's now turn to look at the slightly more complicated plot in Figure 3.6. This plot adds three new components to the mix: facets, multiple layers and statistics. The facets and layers expand the data structure described above: each facet panel in each layer has its own dataset. You can think of this as a 3d array: the panels of the facets form a 2d grid, and the layers extend upwards in the 3rd dimension. In this case the data in the layers is the same, but in general we can plot different datasets on different layers. Table 3.5 shows the first few rows of the data in each facet.

```
qplot(displ, hwy, data=mpg, facets = . ~ year) + geom_smooth()
```

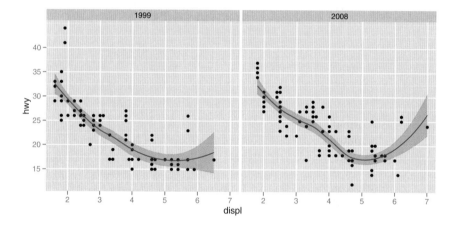

Fig. 3.6: A more complex plot with facets and multiple layers.

The smooth layer is different to the point layer because it doesn't display the raw data, but instead displays a statistical transformation of the data. Specifically, the smooth layer fits a smooth line through the middle of the data. This requires an additional step in the process described above: after mapping the data to aesthetics, the data is passed to a statistical transformation, or **stat**, which manipulates the data in some useful way. In this example, the stat fits the data to a loess smoother, and then returns predictions from evenly spaced points within the range of the data. Other useful stats include 1 and 2d binning, group means, quantile regression and contouring.

As well as adding an additional step to summarise the data, we also need some extra steps when we get to the scales. This is because we now have multiple datasets (for the different facets and layers) and we need to make sure that the scales are the same across all of them. Scaling actually occurs in three parts: transforming, training and mapping. We haven't mentioned

x	y	colour	x	y	colour
1.8	29	4	2.0	31	4
1.8	29	4	2.0	30	4
2.8	26	6	3.1	27	6
2.8	26	6	2.0	28	4
1.8	26	4	2.0	27	4
1.8	25	4	3.1	25	6
2.8	25	6	3.1	25	6
2.8	25	6	3.1	25	6
2.8	24	6	4.2	23	8
5.7	17	8	5.3	20	8

Table 3.5: A 1×2 grid of data frames used for faceting. In general, this structure also has a third dimension for layers, but in this example the data for each layer is the same.

transformation before, but you have probably seen it before in log-log plots. In a log-log plot, the data values are not linearly mapped to position on the plot, but are first log-transformed.

- Scale transformation occurs before statistical transformation so that statistics are computed on the scale-transformed data. This ensures that a plot of $\log(x)$ vs. $\log(y)$ on linear scales looks the same as x vs. y on log scales. There are many different transformations that can be used, including taking square roots, logarithms and reciprocals. See Section 6.4.2 for more details.
- After the statistics are computed, each scale is trained on every dataset from all the layers and facets. The training operation combines the ranges of the individual datasets to get the range of the complete data. Without this step, scales could only make sense locally and we wouldn't be able to overlay different layers because their positions wouldn't line up. Sometimes we do want to vary position scales across facets (but never across layers), and this is described more fully in Section 7.2.3.
- Finally the scales map the data values into aesthetic values. This is a local operation: the variables in each dataset are mapped to their aesthetic values producing a new dataset that can then be rendered by the geoms.

Figure 3.7 illustrates the complete process schematically.

3.5 Components of the layered grammar

In the examples above, we have seen some of the components that make up a plot, data and aesthetic mappings, geometric objects (geoms), statistical transformations (stats), scales and faceting. We have also touched on the coordinate system. One thing we didn't mention is the position adjustment,

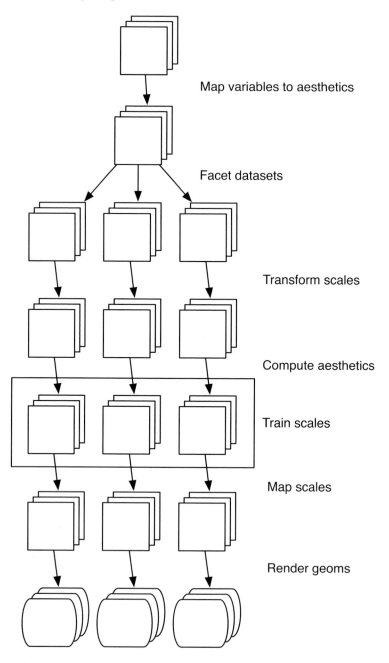

Map variables to aesthetics

Facet datasets

Transform scales

Compute aesthetics

Train scales

Map scales

Render geoms

Fig. 3.7: Schematic description of the plot generation process. Each square represents a layer, and this schematic represents a plot with three layers and three panels. All steps work by transforming individual data frames, except for training scales which doesn't affect the data frame and operates across all datasets simultaneously.

which deals with overlapping graphic objects. Together, the data, mappings, stat, geom and position adjustment form a **layer**. A plot may have multiple layers, as in the example where we overlaid a smoothed line on a scatterplot. All together, the layered grammar defines a plot as the combination of:

- A default dataset and set of mappings from variables to aesthetics.
- One or more layers, each composed of a geometric object, a statistical transformation, and a position adjustment, and optionally, a dataset and aesthetic mappings.
- One scale for each aesthetic mapping.
- A coordinate system.
- The faceting specification.

The following sections describe each of the higher level components more precisely, and point you to the parts of the book where they are documented.

3.5.1 Layers

Layers are responsible for creating the objects that we perceive on the plot. A layer is composed of four parts:

- data and aesthetic mapping,
- a statistical transformation (stat),
- a geometric object (geom)
- and a position adjustment.

The properties of a layer are described in Chapter 4 and how they can be used to visualise data in Chapter 5.

3.5.2 Scales

A **scale** controls the mapping from data to aesthetic attributes, and we need a scale for every aesthetic used on a plot. Each scale operates across all the data in the plot, ensuring a consistent mapping from data to aesthetics. Some scales are illustrated in Figure 3.8.

A scale is a function, and its inverse, along with a set of parameters. For example, the colour gradient scale maps a segment of the real line to a path through a colour space. The parameters of the function define whether the path is linear or curved, which colour space to use (e.g., LUV or RGB), and the colours at the start and end.

The inverse function is used to draw a guide so that you can read values from the graph. Guides are either axes (for position scales) or legends (for everything else). Most mappings have a unique inverse (i.e., the mapping function is one-to-one), but many do not. A unique inverse makes it possible to recover the original data, but this is not always desirable if we want to focus attention on a single aspect.

Chapter 6 describes scales in detail.

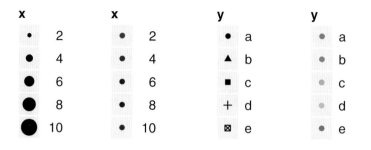

Fig. 3.8: Examples of legends from four different scales. From left to right: continuous variable mapped to size, and to colour, discrete variable mapped to shape, and to colour. The ordering of scales seems upside-down, but this matches the labelling of the y-axis: small values occur at the bottom.

3.5.3 Coordinate system

A coordinate system, or **coord** for short, maps the position of objects onto the plane of the plot. Position is often specified by two coordinates (x, y), but potential could be three or more (although this is not yet implemented in `ggplot2`). The Cartesian coordinate system is the most common coordinate system for two dimensions, while polar coordinates and various map projections are used less frequently.

Coordinate systems affect all position variables simultaneously and differ from scales in that they also change the appearance of the geometric objects. For example, in polar coordinates, bar geoms look like segments of a circle. Additionally, scaling is performed before statistical transformation, while coordinate transformations occur afterward. The consequences of this are shown in Section 7.3.1.

Coordinate systems control how the axes and grid lines are drawn. Figure 3.9 illustrates three different types of coordinate systems. Very little advice is available for drawing these for non-Cartesian coordinate systems, so a lot of work needs to be done to produce polished output. Coordinate systems are described in Section 7.3.

3.5.4 Faceting

There is also another thing that turns out to be sufficiently useful that we should include it in our general framework: faceting, a general case of the conditioned or trellised plots. This makes it easy to create small multiples each showing a different subset of the whole dataset. This is a powerful tool when investigating whether patterns hold across all conditions. The faceting

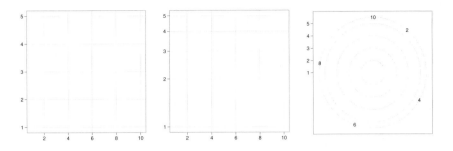

Fig. 3.9: Examples of axes and grid lines for three coordinate systems: Cartesian, semi-log and polar. The polar coordinate system illustrates the difficulties associated with non-Cartesian coordinates: it is hard to draw the axes well.

specification describes which variables should be used to split up the data, and whether position scales should be free or constrained. Faceting is described in Chapter 7.

3.6 Data structures

This grammar is encoded into R data structures in a fairly straightforward way. A plot object is a list with components `data`, `mapping` (the default aesthetic mappings), `layers`, `scales`, `coordinates` and `facet`. The plot object has one other component we haven't discussed yet: `options`. This is used to store the plot-specific theme options described in Chapter 8.

Plots can be created in two ways: all at once with `qplot()`, as shown in the previous chapter, or piece-by-piece with `ggplot()` and layer functions, as described in the next chapter. Once you have a plot object, there are a few things you can do with it:

- Render it on screen, with `print()`. This happens automatically when running interactively, but inside a loop or function, you'll need to `print()` it yourself.
- Render it to disk, with `ggsave()`, described in Section 8.3.
- Briefly describe its structure with `summary()`.
- Save a cached copy of it to disk, with `save()`. This saves a complete copy of the plot object, so you can easily re-create that exact plot with `load()`. Note that data is stored inside the plot, so that if you change the data outside of the plot, and then redraw a saved plot, it will not be updated.

The following code illustrates some of these tools.

```
> p <- qplot(displ, hwy, data = mpg, colour = factor(cyl))
> summary(p)
```

```
data: manufacturer, model, displ, year, cyl, trans,
   drv, cty, hwy, fl, class [234x11]
mapping:  colour = factor(cyl), x = displ, y = hwy
scales:    colour, x, y
faceting: facet_grid(. ~ ., FALSE)
-----------------------------------
geom_point:
stat_identity:
position_identity: (width = NULL, height = NULL)

> # Save plot object to disk
> save(p, file = "plot.rdata")
> # Load from disk
> load("plot.rdata")
> # Save png to disk
> ggsave("plot.png", width = 5, height = 5)
```

Chapter 4

Build a plot layer by layer

4.1 Introduction

Layering is the mechanism by which additional data elements are added to a plot. Each layer can come from a different dataset and have a different aesthetic mapping, allowing us to create plots that could not be generated using `qplot()`, which permits only a single dataset and a single set of aesthetic mappings.

This chapter is mainly a technical description of how layers, geoms, statistics and position adjustments work: how you call and customise them. The next chapter, the "toolbox", describes how you can use different geoms and stats to solve particular visualisation problems. These two chapters are companions, with this chapter explaining the theory and the next chapter explaining the practical aspects of using layers to achieve your graphical goals.

Section 4.2 will teach you how to initialise a plot object by hand, a task that `qplot()` performs for us. The plot is not ready to be displayed until at least one layer is added, as described in Section 4.3. This section first describes the complete layer specification, which helps you see exactly how the components of the grammar are realised in R code, and then shows you the shortcuts that will save you a lot of time. As you have learned in the previous chapter, there are five components of a layer:

- The data, § 4.4, which must be an R data frame, and can be changed after the plot is created.
- A set of aesthetic mappings, § 4.5, which describe how variables in the data are mapped to aesthetic properties of the layer. This section includes a description of how layer settings override the plot defaults, the difference between setting and mapping, and the important group aesthetic.
- The geom, § 4.6, which describes the geometric used to draw the layer. The geom defines the set of available aesthetic properties.
- The stat, § 4.7, which takes the raw data and transforms it in some useful way. The stat returns a data frame with new variables that can also be mapped to aesthetics with a special syntax.

H. Wickham, *ggplot2*, Use R, DOI 10.1007/978-0-387-98141-3_4,
© Springer Science+Business Media, LLC 2009

- The position adjustment, §4.8, which adjusts elements to avoid overplotting.

To conclude, Section 4.9 shows you some plotting techniques that pull together everything you have learned in this chapter to create novel visualisations and to visualise model information along with your data.

4.2 Creating a plot

When we used `qplot()`, it did a lot of things for us: it created a plot object, added layers, and displayed the result, using many default values along the way. To create the plot object ourselves, we use `ggplot()`. This has two arguments: **data** and aesthetic **mapping**. These arguments set up defaults for the plot and can be omitted if you specify data and aesthetics when adding each layer. The data argument needs little explanation: It's the data frame that you want to visualise. You are already familiar with aesthetic mappings from `qplot()`, and the syntax here is quite similar, although you need to wrap the pairs of aesthetic attribute and variable name in the `aes()` function. `aes()` is described more fully in Section 4.5, but it's not very tricky. The following example specifies a default mapping of x to `carat`, y to `price` and colour to `cut`.

```
p <- ggplot(diamonds, aes(carat, price, colour = cut))
```

This plot object cannot be displayed until we add a layer: there is nothing to see!

4.3 Layers

A minimal layer may do nothing more than specify a **geom**, a way of visually representing the data. If we add a point geom to the plot we just created, we create a scatterplot, which can then be rendered.

```
p <- p + layer(geom = "point")
```

Note how we use + to **add** the layer to the plot. This layer uses the plot defaults for data and aesthetic mapping and it uses default values for two optional arguments: the statistical transformation (the stat) and the position adjustment. A more fully specified layer can take any or all of these arguments:

```
layer(geom, geom_params, stat, stat_params, data, mapping,
  position)
```

Here is what a more complicated call looks like. It produces a histogram (a combination of bars and binning) coloured "steelblue" with a bin width of 2:

```
p <- ggplot(diamonds, aes(x = carat))
p <- p + layer(
  geom = "bar",
  geom_params = list(fill = "steelblue"),
  stat = "bin",
  stat_params = list(binwidth = 2)
)
p
```

This layer specification is precise but verbose. We can simplify it by using shortcuts that rely on the fact that every geom is associated with a default statistic and position, and every statistic with a default geom. This means that you only need to specify one of `stat` or `geom` to get a completely specified layer, with parameters passed on to the geom or stat as appropriate. This expression generates the same layer as the full layer command above:

```
geom_histogram(binwidth = 2, fill = "steelblue")
```

All the shortcut functions have the same basic form, beginning with `geom_` or `stat_`:

```
geom_XXX(mapping, data, ..., geom, position)
stat_XXX(mapping, data, ..., stat, position)
```

Their common parameters define the components of the layer:

- **mapping** (optional): A set of aesthetic mappings, specified using the `aes()` function and combined with the plot defaults as described in Section 4.5.
- **data** (optional): A dataset which overrides the default plot dataset. It is most commonly omitted, in which case the layer will use the default plot data. See Section 4.4.
- **...**: Parameters for the geom or stat, such as bin width in the histogram or bandwidth for a loess smoother. You can also use aesthetic properties as parameters. When you do this you **set** the property to a fixed value, not **map** it to a variable in the dataset. The example above showed setting the fill colour of the histogram to "steelblue". See Section 4.5.2 for more examples.
- **geom** or **stat** (optional): You can override the default `stat` for a `geom`, or the default `geom` for a `stat`. This is a text string containing the name of the geom to use. Using the default will give you a standard plot; overriding the default allows you to achieve something more exotic, as shown in Section 4.9.1.
- **position** (optional): Choose a method for adjusting overlapping objects, as described in Section 4.8.

Note that the order of `data` and `mapping` arguments is switched between `ggplot()` and the layer functions. This is because you almost always specify data for the plot, and almost always specify aesthetics—but *not* data—for the layers. We suggest explicitly naming all other arguments rather than relying on positional matching. This makes the code more readable and is the style followed in this book.

Layers can be added to plots created with `ggplot()` or `qplot()`. Remember, behind the scenes, `qplot()` is doing exactly the same thing: it creates a plot object and then adds layers. The following example shows the equivalence between these two ways of making plots.

```
ggplot(msleep, aes(sleep_rem / sleep_total, awake)) +
  geom_point()
# which is equivalent to
qplot(sleep_rem / sleep_total, awake, data = msleep)

# You can add layers to qplot too:
qplot(sleep_rem / sleep_total, awake, data = msleep) +
  geom_smooth()
# This is equivalent to
qplot(sleep_rem / sleep_total, awake, data = msleep,
  geom = c("point", "smooth"))
# or
ggplot(msleep, aes(sleep_rem / sleep_total, awake)) +
  geom_point() + geom_smooth()
```

You've seen that plot objects can be stored as variables. The summary function can be helpful for inspecting the structure of a plot without plotting it, as seen in the following example. The summary shows information about the plot defaults, and then each layer. You will learn about scales and faceting in Chapters 6 and 7.

```
> p <- ggplot(msleep, aes(sleep_rem / sleep_total, awake))
> summary(p)
data: name, genus, vore, order, conservation,
  sleep_total, sleep_rem, sleep_cycle, awake,
  brainwt, bodywt [83x11]
mapping:  x = sleep_rem/sleep_total, y = awake
scales:   x, y
faceting: facet_grid(. ~ ., FALSE)
>
> p <- p + geom_point()
> summary(p)
data: name, genus, vore, order, conservation,
  sleep_total, sleep_rem, sleep_cycle, awake,
  brainwt, bodywt [83x11]
mapping:  x = sleep_rem/sleep_total, y = awake
scales:   x, y
```

```
faceting: facet_grid(. ~ ., FALSE)
-----------------------------------
geom_point: na.rm = FALSE
stat_identity:
position_identity: (width = NULL, height = NULL)
```

Layers are regular R objects and so can be stored as variables, making it easy to write clean code that reduces duplication. For example, a set of plots can be initialised using different data then enhanced with the same layer. If you later decide to change that layer, you only need to do so in one place. The following shows a simple example, where we create a layer that displays a translucent thick blue line of best fit.

```
bestfit <- geom_smooth(method = "lm", se = F,
  colour = alpha("steelblue", 0.5), size = 2)
qplot(sleep_rem, sleep_total, data = msleep) + bestfit
qplot(awake, brainwt, data = msleep, log = "y") + bestfit
qplot(bodywt, brainwt, data = msleep, log = "xy") + bestfit
```

The following sections describe data and mappings in more detail, then go on to describe the available geoms, stats and position adjustments.

4.4 Data

The restriction on the data is simple: it must be a data frame. This is restrictive, and unlike other graphics packages in R. Lattice functions can take an optional data frame or use vectors directly from the global environment. Base methods often work with vectors, data frames or other R objects. However, there are good reasons for this restriction. Your data is very important, and it's better to be explicit about exactly what is done with it. It also allows a cleaner separation of concerns so that ggplot2 deals only with plotting data, not wrangling it into different forms, for which you might find the plyr or reshape packages helpful. A single data frame is also easier to save than a multitude of vectors, which means it's easier to reproduce your results or send your data to someone else.

This restriction also makes it very easy to produce the same plot for different data: you just change the data frame. You can replace the old dataset with %+%, as shown in the following example. (You might expect that this would use + like all the other components, but unfortunately due to a restriction in R this is not possible.) Swapping out the data makes it easy to experiment with imputation schemes or model fits, as shown in Section 4.9.3.

```
p <- ggplot(mtcars, aes(mpg, wt, colour = cyl)) + geom_point()
p
```

```
mtcars <- transform(mtcars, mpg = mpg ^ 2)
p %+% mtcars
```

Any change of values or dimensions is legitimate. However, if a variable changes from discrete to continuous (or vice versa), you will need to change the default scales, as described in Section 6.3.

It is not necessary to specify a default dataset except when using faceting; faceting is a global operation (i.e., it works on all layers) and it needs to have a base dataset which defines the set of facets for all datasets. See Section 7.2.4 for more details. If the default dataset is omitted, every layer must supply its own data.

The data is stored in the plot object as a copy, not a reference. This has two important consequences: if your data changes, the plot will not; and ggplot2 objects are entirely self-contained so that they can be save()d to disk and later load()ed and plotted without needing anything else from that session.

4.5 Aesthetic mappings

To describe the way that variables in the data are mapped to things that we can perceive on the plot (the "aesthetics"), we use the aes function. The aes function takes a list of aesthetic-variable pairs like these:

```
aes(x = weight, y = height, colour = age)
```

Here we are mapping x-position to weight, y-position to height and colour to age. The first two arguments can be left without names, in which case they correspond to the x and y variables. This matches the way that qplot() is normally used. You should never refer to variables outside of the dataset (e.g., with diamonds$carat), as this makes it impossible to encapsulate all of the data needed for plotting in a single object.

```
aes(weight, height, colour = sqrt(age))
```

Note that functions of variables can be used.

Any variable in an aes() specification must be contained inside the plot or layer data. This is one of the ways in which ggplot2 objects are guaranteed to be entirely self-contained, so that they can be stored and re-used.

4.5.1 Plots and layers

The default aesthetic mappings can be set when the plot is initialised or modified later using +, as in this example:

```
> p <- ggplot(mtcars)
> summary(p)
data: mpg, cyl, disp, hp, drat, wt, qsec, vs, am,
```

```
  gear, carb [32x11]
faceting: facet_grid(. ~ ., FALSE)
>
> p <- p + aes(wt, hp)
> summary(p)
data: mpg, cyl, disp, hp, drat, wt, qsec, vs, am,
  gear, carb [32x11]
mapping:  x = wt, y = hp
scales:   list(), list()
faceting: facet_grid(. ~ ., FALSE)
```

One reason you might want to do this is shown in Section 4.9.3. We have seen several examples of using the default mapping when adding a layer to a plot:

```
> p <- ggplot(mtcars, aes(x = mpg, y = wt))
> p + geom_point()
```

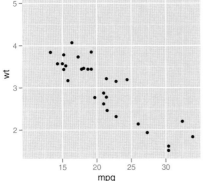

The default mappings in the plot p can be extended or overridden in the layers, as with the following code. The results are shown in Figure 4.1.

```
p + geom_point(aes(colour = factor(cyl)))
p + geom_point(aes(y = disp))
```

The rules are summarised in Table 4.1. Aesthetic mappings specified in a layer affect only that layer. For that reason, unless you modify the default scales, axis labels and legend titles will be based on the plot defaults. The way to change these is described in Section 6.5.

4.5.2 Setting vs. mapping

Instead of mapping an aesthetic property to a variable, you can set it to a single value by specifying it in the layer parameters. Aesthetics can vary

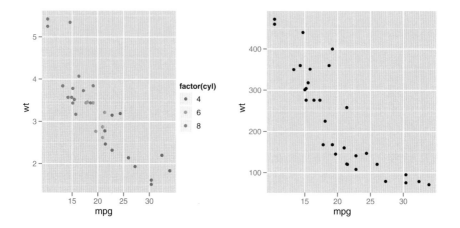

Fig. 4.1: Overriding aesthetics. (Left) Overriding colour with `factor(cyl)` and (right) overriding y-position with `disp`

Operation	Layer aesthetics	Result
Add	`aes(colour = cyl)`	`aes(mpg, wt, colour = cyl)`
Override	`aes(y = disp)`	`aes(mpg, disp)`
Remove	`aes(y = NULL)`	`aes(mpg)`

Table 4.1: Rules for combining layer mappings with the default mapping of `aes(mpg, wt)`. Layer aesthetics can add to, override, and remove the default mappings.

for each observation being plotted, while parameters do not. We **map** an aesthetic to a variable (e.g., (aes(colour = cut))) or **set** it to a constant (e.g., `colour = "red"`). For example, the following layer sets the colour of the points, using the colour parameter of the layer:

```
p <- ggplot(mtcars, aes(mpg, wt))
p + geom_point(colour = "darkblue")
```

This sets the point colour to be dark blue instead of black. This is quite different than

```
p + geom_point(aes(colour = "darkblue"))
```

This **maps** (not sets) the colour to the value "darkblue". This effectively creates a new variable containing only the value "darkblue" and then maps colour to that new variable. Because this value is discrete, the default colour scale uses evenly spaced colours on the colour wheel, and since there is only one value this colour is pinkish. The difference between setting and mapping is illustrated in Figure 4.2.

With `qplot()`, you can do the same thing by putting the value inside of
`I()`, e.g., `colour = I("darkblue")`. Chapter B describes how values should
be specified for the various aesthetics.

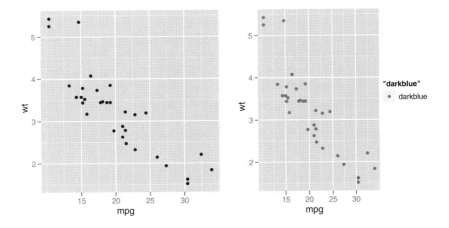

Fig. 4.2: The difference between (left) setting colour to `"darkblue"` and (right)
mapping colour to `"darkblue"`. When `"darkblue"` is mapped to colour, it is treated
as a regular value and scaled with the default colour scale. This results in pinkish
points and a legend.

4.5.3 Grouping

In `ggplot2`, geoms can be roughly divided into individual and collective geoms.
An individual geom has a distinctive graphical object for each row in the data
frame. For example, the point geom has a single point for each observation.
On the other hand, collective geoms represent multiple observations. This may
be a result of a statistical summary, or may be fundamental to the display
of the geom, as with polygons. Lines and paths fall somewhere in between:
each overall line is composed of a set of straight segments, but each segment
represents two points. How do we control which observations go in which
individual graphical element? This is the job of the `group` aesthetic.

By default, the `group` is set to the interaction of all discrete variables in the
plot. This often partitions the data correctly, but when it does not, or when
no discrete variable is used in the plot, you will need to explicitly define the
grouping structure, by mapping group to a variable that has a different value
for each group. The `interaction()` function is useful if a single pre-existing
variable doesn't cleanly separate groups, but a combination does.

There are three common cases where the default is not enough, and we
will consider each one below. In the following examples, we will use a simple

longitudinal dataset, `Oxboys`, from the `nlme` package. It records the heights (height) and centered ages (age) of 26 boys (Subject), measured on nine occasions (Occasion).

Multiple groups, one aesthetic.

In many situations, you want to separate your data into groups, but render them in the same way. When looking at the data in aggregate you want to be able to distinguish individual subjects, but not identify them. This is common in longitudinal studies with many subjects, where the plots are often descriptively called spaghetti plots.

The first plot in Figure 4.3 shows a set of time series plots, one for each boy. You can see the separate growth trajectories for each boy, but there is no way to see which boy belongs to which trajectory. This plot was generated with:

```
p <- ggplot(Oxboys, aes(age, height, group = Subject)) +
  geom_line()
```

We specified the Subject as the grouping variable to get a line for each boy. The second plot in the figure shows the result of leaving this out: we get a single line which passes through every point. This is not very useful! Line plots with an incorrect grouping specification typically have this characteristic appearance.

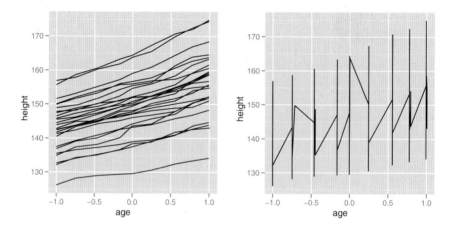

Fig. 4.3: (Left) Correctly specifying `group = Subject` produces one line per subject. (Right) A single line connects all observations. This pattern is characteristic of an incorrect grouping aesthetic, and is what we see if the group aesthetic is omitted, which in this case is equivalent to `group = 1`.

Different groups on different layers.

Sometimes we want to plot summaries based on different levels of aggregation. Different layers might have different group aesthetics, so that some display individual level data while others display summaries of larger groups.

Building on the previous example, suppose we want to add a single smooth line to the plot just created, based on the ages and heights of *all* the boys. If we use the same grouping for the smooth that we used for the line, we get the first plot in Figure 4.4.

```
p + geom_smooth(aes(group = Subject), method="lm", se = F)
```

This is not what we wanted; we have inadvertently added a smoothed line for each boy. This new layer needs a different group aesthetic, `group = 1`, so that the new line will be based on all the data, as shown in the second plot in the figure. The modified layer looks like this:

```
p + geom_smooth(aes(group = 1), method="lm", size = 2, se = F)
```

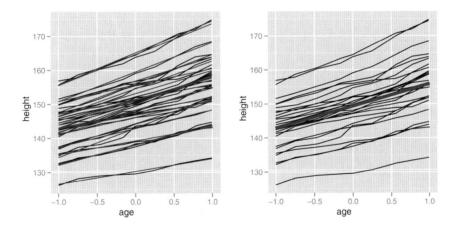

Fig. 4.4: Adding smooths to the Oxboys data. (Left) Using the same grouping as the lines results in a line of best fit for each boy. (Right) Using `aes(group = 1)` in the smooth layer fits a single line of best fit across all boys.

Note how we stored the first plot in the variable **p**, so we could experiment with the code to generate the second layer without having to re-enter any of the code for the first layer. This is a useful time-saving technique, and is expanded upon in Chapter 10.

Overriding the default grouping.

The plot has a discrete scale but you want to draw lines that connect *across* groups. This is the strategy used in interaction plots, profile plots, and parallel coordinate plots, among others. For example, we draw boxplots of height at each measurement occasion, as shown in the first figure in Figure 4.5:

```
boysbox <- ggplot(Oxboys, aes(Occasion, height)) + geom_boxplot()
```

There is no need to specify the group aesthetic here; the default grouping works because occasion is a discrete variable. To overlay individual trajectories we again need to override the default grouping for that layer with `aes(group = Subject)`, as shown in the second plot in the figure.

```
boysbox + geom_line(aes(group = Subject), colour = "#3366FF")
```

We change the line colour in the second layer to make them distinct from the boxes. This is another example of setting an aesthetic to a fixed value. The colour is a rendering attribute, which has no corresponding variable in the data.

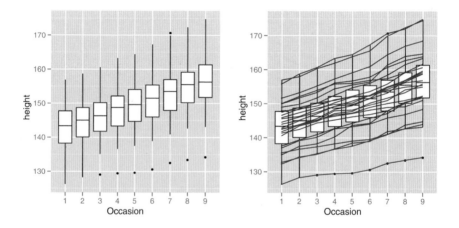

Fig. 4.5: (Left) If boxplots are used to look at the distribution of heights at each occasion (a discrete variable), the default grouping works correctly. (Right) If trajectories of individual boys are overlaid with geom_line(), then aes(group = Subject) is needed for the new layer.

4.5.4 Matching aesthetics to graphic objects

Another important issue with collective geom is how the aesthetics of the individual observations are mapped to the aesthetics of the complete entity. For

individual geoms, this isn't a problem, because each observation is represented
by a single graphical element. However, high data densities can make it difficult
(or impossible) to distinguish between individual points and in some sense the
point geom becomes a collective geom, a single blob of points.

Lines and paths operate on an off-by-one principle: there is one more
observation than line segment, and so the aesthetic for the first observation is
used for the first segment, the second observation for the second segment and
so on. This means that the aesthetic for the last observation is not used, as
shown in Figure 4.6. An additional limitation for paths and lines is that that
line type must be constant over each individual line, in R there is no way to
draw a joined up line which has varying line type.

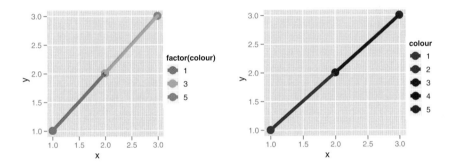

Fig. 4.6: For lines and paths, the aesthetics of the line segment are determined by
the aesthetic of the beginning observation. If colour is categorical (left) there is
no meaningful way to interpolate between adjacent colours. If colour is continuous
(right), there is, but this is not done by default.

You could imagine a more complicated system where segments smoothly
blend from one aesthetic to another. This would work for continuous variables
like size or colour, but not for line type, and is not used in ggplot2. If this is
the behaviour you want, you can perform the linear interpolation yourself, as
shown below.

```
> xgrid <- with(df, seq(min(x), max(x), length = 50))
> interp <- data.frame(
+   x = xgrid,
+   y = approx(df$x, df$y, xout = xgrid)$y,
+   colour = approx(df$x, df$colour, xout = xgrid)$y
+ )
> qplot(x, y, data = df, colour = colour, size = I(5)) +
+   geom_line(data = interp, size = 2)
```

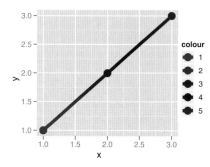

For all other collective geoms, like polygons, the aesthetics from the individual components are only used if they are all the same, otherwise the default value is used. This makes sense for fill as it is a property of the entire object: it doesn't make sense to think about having a different fill colour for each point on the border of the polygon.

These issues are most relevant when mapping aesthetics to continuous variable, because, as described above, when you introduce a mapping to a discrete variable, it will by default split apart collective geoms into smaller pieces. This works particularly well for bar and area plots, because stacking the individual pieces produces the same shape as the original ungrouped data. This is illustrated in Figure 4.7.

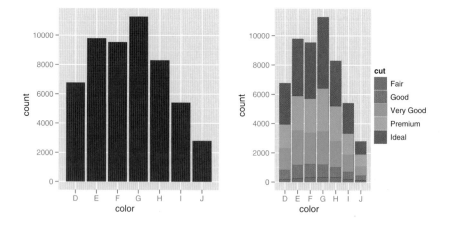

Fig. 4.7: Splitting apart a bar chart (left) produces a plot (right) that has the same outline as the original.

4.6 Geoms

Geometric objects, or **geom**s for short, perform the actual rendering of the layer, control the type of plot that you create. For example, using a point geom will create a scatterplot, while using a line geom will create a line plot. Table 4.2 lists all of the geoms available in `ggplot2`.

Each geom has a set of aesthetics that it understands, and a set that are required for drawing. For example, a point requires x and y position, and understands colour, size and shape aesthetics. A bar requires height (`ymax`), and understands width, border colour and fill colour. These are listed for all geoms in Table 4.3.

Some geoms differ primarily in the way that they are parameterised. For example, the tile geom is specified in terms of the location of its centre and its height and width, while the rect geom is parameterised in terms of its top (`ymax`), bottom (`ymin`), left (`xmin`) and right (`right`) positions. Internally, the rect geom is described as a polygon, and it is parameters are the locations of the four corners. This is useful for non-Cartesian coordinate systems, as you will learn in Chapter 7.

Every geom has a default statistic, and every statistic a default geom. For example, the bin statistic defaults to using the bar geom to produce a histogram. These defaults are listed in Table 4.3. Overriding these defaults will still produce valid plots, but they may violate graphical conventions. See examples in Section 4.9.1.

4.7 Stat

A statistical transformation, or **stat**, transforms the data, typically by summarising it in some manner. For example, a useful stat is the smoother, which calculates the mean of y, conditional on x, subject to some restriction that ensures smoothness. All currently available stats are listed in Table 4.4. To make sense in a graphic context a stat must be location-scale invariant: $f(x + a) = f(x) + a$ and $f(b \cdot x) = b \cdot f(x)$. This ensures that the transformation stays the same when you change the scales of the plot.

A stat takes a dataset as input and returns a dataset as output, and so a stat can add new variables to the original dataset. It is possible to map aesthetics to these new variables. For example, `stat_bin`, the statistic used to make histograms, produces the following variables:

- count, the number of observations in each bin
- density, the density of observations in each bin (percentage of total / bar width)
- x, the centre of the bin

These generated variables can be used instead of the variables present in the original dataset. For example, the default histogram geom assigns the

Name	Description
abline	Line, specified by slope and intercept
area	Area plots
bar	Bars, rectangles with bases on y-axis
blank	Blank, draws nothing
boxplot	Box-and-whisker plot
contour	Display contours of a 3d surface in 2d
crossbar	Hollow bar with middle indicated by horizontal line
density	Display a smooth density estimate
density_2d	Contours from a 2d density estimate
errorbar	Error bars
histogram	Histogram
hline	Line, horizontal
interval	Base for all interval (range) geoms
jitter	Points, jittered to reduce overplotting
line	Connect observations, in order of x value
linerange	An interval represented by a vertical line
path	Connect observations, in original order
point	Points, as for a scatterplot
pointrange	An interval represented by a vertical line, with a point in the middle
polygon	Polygon, a filled path
quantile	Add quantile lines from a quantile regression
ribbon	Ribbons, y range with continuous x values
rug	Marginal rug plots
segment	Single line segments
smooth	Add a smoothed condition mean
step	Connect observations by stairs
text	Textual annotations
tile	Tile plot as densely as possible, assuming that every tile is the same size
vline	Line, vertical

Table 4.2: Geoms in `ggplot2`

Name	Default stat	Aesthetics
abline	abline	colour, linetype, size
area	identity	colour, fill, linetype, size, **x**, **y**
bar	bin	colour, fill, linetype, size, weight, **x**
bin2d	bin2d	colour, fill, linetype, size, weight, **xmax**, **xmin**, **ymax**, **ymin**
blank	identity	
boxplot	boxplot	colour, fill, **lower**, **middle**, size, **upper**, weight, **x**, **ymax**, **ymin**
contour	contour	colour, linetype, size, weight, **x**, **y**
crossbar	identity	colour, fill, linetype, size, **x**, **y**, **ymax**, **ymin**
density	density	colour, fill, linetype, size, weight, **x**, **y**
density2d	density2d	colour, linetype, size, weight, **x**, **y**
errorbar	identity	colour, linetype, size, width, **x**, **ymax**, **ymin**
freqpoly	bin	colour, linetype, size
hex	binhex	colour, fill, size, **x**, **y**
histogram	bin	colour, fill, linetype, size, weight, **x**
hline	hline	colour, linetype, size
jitter	identity	colour, fill, shape, size, **x**, **y**
line	identity	colour, linetype, size, **x**, **y**
linerange	identity	colour, linetype, size, **x**, **ymax**, **ymin**
path	identity	colour, linetype, size, **x**, **y**
point	identity	colour, fill, shape, size, **x**, **y**
pointrange	identity	colour, fill, linetype, shape, size, **x**, **y**, **ymax**, **ymin**
polygon	identity	colour, fill, linetype, size, **x**, **y**
quantile	quantile	colour, linetype, size, weight, **x**, **y**
rect	identity	colour, fill, linetype, size, **xmax**, **xmin**, **ymax**, **ymin**
ribbon	identity	colour, fill, linetype, size, **x**, **ymax**, **ymin**
rug	identity	colour, linetype, size
segment	identity	colour, linetype, size, **x**, **xend**, **y**, **yend**
smooth	smooth	alpha, colour, fill, linetype, size, weight, **x**, **y**
step	identity	colour, linetype, size, **x**, **y**
text	identity	angle, colour, hjust, **label**, size, vjust, **x**, **y**
tile	identity	colour, fill, linetype, size, **x**, **y**
vline	vline	colour, linetype, size

Table 4.3: Default statistics and aesthetics. Emboldened aesthetics are required.

Name	Description
bin	Bin data
boxplot	Calculate components of box-and-whisker plot
contour	Contours of 3d data
density	Density estimation, 1d
density_2d	Density estimation, 2d
function	Superimpose a function
identity	Don't transform data
qq	Calculation for quantile-quantile plot
quantile	Continuous quantiles
smooth	Add a smoother
spoke	Convert angle and radius to xend and yend
step	Create stair steps
sum	Sum unique values. Useful for overplotting on scatter-plots
summary	Summarise y values at every unique x
unique	Remove duplicates

Table 4.4: Stats in `ggplot2`

height of the bars to the number of observations (count), but if you'd prefer a more traditional histogram, you can use the density (density). The following example shows a density histogram of carat from the diamonds dataset.

```
> ggplot(diamonds, aes(carat)) +
+   geom_histogram(aes(y = ..density..), binwidth = 0.1)
```

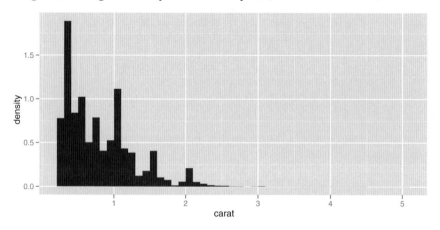

The names of generated variables must be surrounded with .. when used. This prevents confusion in case the original dataset includes a variable with the same name as a generated variable, and it makes it clear to any later reader of the code that this variable was generated by a stat. Each statistic lists the variables that it creates in its documentation.

The syntax to produce this plot with `qplot()` is very similar:

```
qplot(carat, ..density.., data = diamonds, geom="histogram",
  binwidth = 0.1)
```

4.8 Position adjustments

Position adjustments apply minor tweaks to the position of elements within a layer. Table 4.5 lists all of the position adjustments available within ggplot2. Position adjustments are normally used with discrete data. Continuous data typically doesn't overlap exactly, and when it does (because of high data density) minor adjustments, like jittering, are usually insufficient to fix the problem.

Adjustment	Description
dodge	Adjust position by dodging overlaps to the side
fill	Stack overlapping objects and standardise have equal height
identity	Don't adjust position
jitter	Jitter points to avoid overplotting
stack	Stack overlapping objects on top of one another

Table 4.5: The five position adjustments.

The different types of adjustment are best illustrated with a bar chart. Figure 4.8 shows stacking, filling and dodging. Stacking puts bars on the same x on top of one another; filling does the same, but normalises height to 1; and dodging places the bars side-by-side. Dodging is rather similar to faceting, and the advantages and disadvantages of each method are described in Section 7.2.6. For these operations to work, each bar must have the same width and not overlap with any others. The identity adjustment (i.e., do nothing) doesn't make much sense for bars, but is shown in Figure 4.9 along with a line plot of the same data for reference.

4.9 Pulling it all together

Once you have become comfortable with combining layers, you will be able to create graphics that are both intricate and useful. The following examples demonstrate some of the ways to use the capabilities of layers that have been introduced in this chapter. These are just to get you started. You are limited only by your imagination!

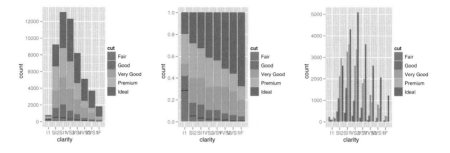

Fig. 4.8: Three position adjustments applied to a bar chart. From left to right, stacking, filling and dodging.

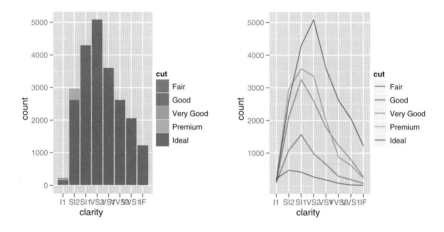

Fig. 4.9: The identity positon adjustment is not useful for bars, (left) because each bar obscures the bars behind. (Right) It is useful for lines, however, because lines do not have the same problem.

4.9.1 Combining geoms and stats

By connecting geoms with different statistics, you can easily create new graphics. Figure 4.10 shows three variations on a histogram. They all use the same statistical transformation underlying a histogram (the bin stat), but use different geoms to display the results: the area geom, the point geom and the tile geom.

```
d <- ggplot(diamonds, aes(carat)) + xlim(0, 3)
d + stat_bin(aes(ymax = ..count..), binwidth = 0.1, geom = "area")
d + stat_bin(
  aes(size = ..density..), binwidth = 0.1,
  geom = "point", position="identity"
```

```
)
d + stat_bin(
  aes(y = 1, fill = ..count..), binwidth = 0.1,
  geom = "tile", position="identity"
)
```

(The use of xlim() will be discussed in Section 6.4.2, in the presentation of the use of scales and axes, but you can already guess that it is used here to set the limits of the horizontal axis.)

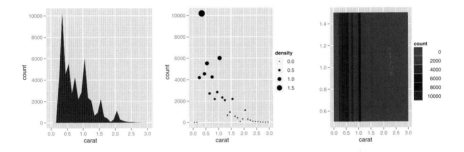

Fig. 4.10: Three variations on the histogram. (Left) A frequency polygon; (middle) a scatterplot with both size and height mapped to frequency; (right) a heatmap representing frequency with colour.

A number of the geoms available in ggplot2 were derived from other geoms in a process like the one just described, starting with an existing geom and making a few changes in the default aesthetics or stat. For example, the jitter geom is simply the point geom with the default position adjustment set to jitter. Once it becomes clear that a particular variant is going to be used a lot or used in a very different context, it makes sense to create a new geom. Table 4.6 lists these "aliased" geoms.

Aliased geom	Base geom	Changes in default
area	ribbon	aes(min = 0, max = y), position = "stack"
density	area	stat = "density"
freqpoly	line	stat = "bin"
histogram	bar	stat = "bin"
jitter	point	position = "jitter"
quantile	line	stat = "quantile"
smooth	ribbon	stat = "smooth"

Table 4.6: Geoms that were created by modifying the defaults of another geom.

4.9.2 Displaying precomputed statistics

If you have data which has already been summarised, and you just want to use it, you'll need to use `stat_identity()`, which leaves the data unchanged, and then map the appropriate variables to the appropriate aesthetics.

4.9.3 Varying aesthetics and data

One of the more powerful capabilities of `ggplot2` is the ability to plot different datasets on different layers. This may seem strange: Why would you want to plot different data on the same plot? In practice, you often have related datasets that should be shown together. A very common example is supplementing the data with predictions from a model. While the smooth geom can add a wide range of different smooths to your plot, it is no substitute for an external quantitative model that summarises your understanding of the data.

Let's look again at the `Oxboys` dataset which was used in Section 4.5.3. In Figure 4.4, we showed linear fits for individual boys (left) and for the whole group (right). Neither model is particularly appropriate: The group model ignores the within-subject correlation and the individual model doesn't use information about the typical growth pattern to more accurately predict individuals. In practice we might use a mixed model to do better. This section explores how we can combine the output from this more sophisticated model with the original data to gain more insight into both the data and the model.

First we'll load the `nlme` package, and fit a model with varying intercepts and slopes. (Exploring the fit of individual models shows that this is a reasonable first pass.) We'll also create a plot to use as a template. This regenerates the first plot in Figure 4.3, but we're not going to render it until we've added data from the model.

```
> require(nlme, quiet = TRUE, warn.conflicts = FALSE)
> model <- lme(height ~ age, data = Oxboys,
+   random = ~ 1 + age | Subject)
> oplot <- ggplot(Oxboys, aes(age, height, group = Subject)) +
+   geom_line()
```

Next we'll compare the predicted trajectories to the actual trajectories. We do this by building up a grid that contains all combinations of ages and subjects. This is overkill for this simple linear case, where we only need two values of age to draw the predicted straight line, but we show it here because it is necessary when the model is more complex. Next we add the predictions from the model back into this dataset, as a variable called **height**.

```
> age_grid <- seq(-1, 1, length = 10)
> subjects <- unique(Oxboys$Subject)
>
> preds <- expand.grid(age = age_grid, Subject = subjects)
> preds$height <- predict(model, preds)
```

Once we have the predictions we can display them along with the original data. Because we have used the same variable names as the original Oxboys dataset, and we want the same group aesthetic, we don't need to specify any aesthetics; we only need to override the default dataset. We also set two aesthetic parameters to make it a bit easier to compare the predictions to the actual values.

```
> oplot + geom_line(data = preds, colour = "#3366FF", size= 0.4)
```

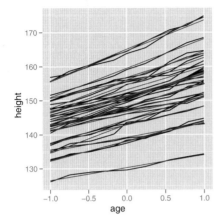

It seems that the model does a good job of capturing the high-level structure of the data, but it's hard to see the details: plots of longitudinal data are often called spaghetti plots, and with good reason. Another way to compare the model to the data is to look at residuals, so let's do that. We add the predictions from the model to the original data (fitted), calculate residuals (resid), and add the residuals as well. The next plot is a little more complicated: We update the plot dataset (recall the use of %+% to update the default data), change the default y aesthetic to resid, and add a smooth line for all observations.

```
> Oxboys$fitted <- predict(model)
> Oxboys$resid <- with(Oxboys, fitted - height)
>
> oplot %+% Oxboys + aes(y = resid) + geom_smooth(aes(group=1))
```

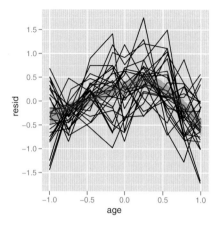

The smooth line makes it evident that the residuals are not random, showing a deficiency in the model. We add a quadratic term, refit the model, recalculate predictions and residuals, and replot. There is now less evidence of model inadequacy.

```
> model2 <- update(model, height ~ age + I(age ^ 2))
> Oxboys$fitted2 <- predict(model2)
> Oxboys$resid2 <- with(Oxboys, fitted2 - height)
>
> oplot %+% Oxboys + aes(y = resid2) + geom_smooth(aes(group=1))
```

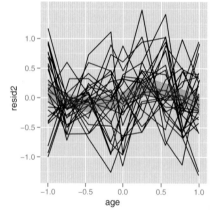

Notice how easily we were able to modify the plot object. We updated the data and replotted twice without needing to reinitialise `oplot`. Layering in `ggplot2` is designed to work well with the iterative process of fitting and evaluating models.

Chapter 5

Toolbox

5.1 Introduction

The layered structure of `ggplot2` encourages you to design and construct graphics in a structured manner. You have learned what a layer is and how to add one to your graphic, but not what geoms and statistics are available to help you build revealing plots. This chapter lists some of the many geoms and stats included in `ggplot2`, broken down by their purpose. This chapter will provide a good overview of the available options, but it does not describe each geom and stat in detail. For more information about individual geoms, along with many more examples illustrating their use, see the online and electronic documentation. You may also want to consult the documentation to learn more about the datasets used in this chapter.

This chapter is broken up into the following sections, each of which deals with a particular graphical challenge. This is not an exhaustive or exclusive categorisation, and there are many other possible ways to break up graphics into different categories. Each geom can be used for many different purposes, especially if you are creative. However, this breakdown should cover many common tasks and help you learn about some of the possibilities.

- Basic plot types, § 5.3, to produce common, "named" graphics like scatterplots and line charts
- Displaying distributions, § 5.4, continuous and discrete, 1d and 2d, joint and conditional
- Dealing with overplotting in scatterplots, § 5.5, a challenge with large datasets
- Surface plots, § 5.6, display 3d surfaces in 2d.
- Statistical summaries, § 5.9, display informative data summaries
- Drawing maps, § 5.7
- Revealing uncertainty and error, § 5.8, with various 1d and 2d intervals
- Annotating a plot, § 5.10, to label, describe and explain with supplemental information
- Weighted data, § 5.11

H. Wickham, *ggplot2*, Use R, DOI 10.1007/978-0-387-98141-3_5,
© Springer Science+Business Media, LLC 2009

The examples in this section use a mixture of `ggplot()` and `qplot()` calls, reflecting real-life use. If you need a reminder on how to translate between the two, see Appendix A.2. The examples do not go into much depth, but hopefully if you flick through this chapter, you'll be able to see a plot that looks like the one you're trying to create.

5.2 Overall layering strategy

It is useful to think about the purpose of each layer before it is added. In general, there are three purposes for a layer:

- To display the **data**. We plot the raw data for many reasons, relying on our skills at pattern detection to spot gross structure, local structure, and outliers. This layer appears on virtually every graphic. In the earliest stages of data exploration, it is often the only layer.
- To display a statistical **summary** of the data. As we develop and explore models of the data, it is useful to display model predictions in the context of the data. We learn from the data summaries and we evaluate the model. Showing the data helps us improve the model, and showing the model helps reveal subtleties of the data that we might otherwise miss. Summaries are usually drawn on top of the data.

  If you review the examples in the preceding chapter, you'll see many examples of plots of data with an added layer displaying a statistical summary.
- To add additional **metadata**, context and annotations. A metadata layer displays background context or annotations that help to give meaning to the raw data. Metadata can be useful in the background and foreground. A map is often used as a background layer with spatial data. Background metadata should be rendered so that it doesn't interfere with your perception of the data, so is usually displayed underneath the data and formatted so that it is minimally perceptible. That is, if you concentrate on it, you can see it with ease, but it doesn't jump out at you when you are casually browsing the plot.

  Other metadata is used to highlight important features of the data. If you have added explanatory labels to a couple of inflection points or outliers, then you want to render them so that they pop out at the viewer. In that case, you want this to be the very last layer drawn.

5.3 Basic plot types

These geoms are the fundamental building blocks of `ggplot2`. They are useful in their own right, but also to construct more complex geoms. Most of these geoms are associated with a named plot: when that geom is used by itself in a plot, that plot has a special name.

Each of these geoms is two dimensional and requires both x and y aesthetics. All understand `colour` and `size` aesthetics, and the filled geoms (bar, tile and polygon) also understand `fill`. The point geom uses `shape` and line and path geoms understand `linetype`. The geoms are used for displaying data, summaries computed elsewhere, and metadata.

- `geom_area()` draws an **area plot**, which is a line plot filled to the y-axis (filled lines). Multiple groups will be stacked on top of each other.
- `geom_bar(stat = "identity")()` makes a **barchart**. We need `stat = "identity"` because the default stat automatically counts values (so is essentially a 1d geom, see § 5.4). The identity stat leaves the data unchanged.

  By default, multiple bars in the same location will be stacked on top of one another.
- `geom_line()` makes a **line plot**. The `group` aesthetic determines which observations are connected; see Section 4.5.3 for more details. `geom_path` is similar to a `geom_line`, but lines are connected in the order they appear in the data, not from left to right.
- `geom_point()` produces a **scatterplot**.
- `geom_polygon()` draws polygons, which are filled paths. Each vertex of the polygon requires a separate row in the data. It is often useful to merge a data frame of polygon coordinates with the data just prior to plotting. Section 5.7 illustrates this concept in more detail for map data.
- `geom_text()` adds labels at the specified points. This is the only geom in this group that requires another aesthetic: `label`. It also has optional aesthetics `hjust` and `vjust` that control the horizontal and vertical position of the text; and `angle` which controls the rotation of the text. See Appendex B for more details.
- `geom_tile()` makes a image plot or level plot. The tiles form a regular tessellation of the plane and typically have the `fill` aesthetic mapped to another variable.

Each of these geoms is illustrated in Figure 5.1, created with the code below.

```
df <- data.frame(
  x = c(3, 1, 5),
  y = c(2, 4, 6),
  label = c("a","b","c")
)
p <- ggplot(df, aes(x, y, label = label)) +
  xlab(NULL) + ylab(NULL)
p + geom_point() + opts(title = "geom_point")
p + geom_bar(stat="identity") +
  opts(title = "geom_bar(stat=\"identity\")")
```

```
p + geom_line() + opts(title = "geom_line")
p + geom_area() + opts(title = "geom_area")
p + geom_path() + opts(title = "geom_path")
p + geom_text() + opts(title = "geom_text")
p + geom_tile() + opts(title = "geom_tile")
p + geom_polygon() + opts(title = "geom_polygon")
```

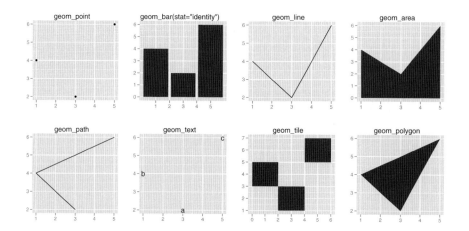

Fig. 5.1: The basic geoms applied to the same data. Many give rise to to named plots (from top left to bottom right): scatterplot, bar chart, line chart, area chart, path plot, labelled scatterplot, image/level plot and polygon plot. Observe the different axis ranges for the bar, area and tile plots: these geoms take up space outside the range of the data, and so push the axes out.

5.4 Displaying distributions

There are a number of geoms that can be used to display distributions, depending on the dimensionality of the distribution, whether it is continuous or discrete, and whether you are interested in conditional or joint distribution.

For 1d continuous distributions the most important geom is the histogram. Figure 5.2 uses the histogram to display the distribution of diamond depth. It is important to experiment with bin placement to find a revealing view. You can change the binwidth, or specify the exact location of the breaks.

If you want to compare the distribution between groups, you have a few options: create small multiples of the histogram, facets = . ˜ var; use a frequency polygon, geom = "freqpoly"; or create a conditional density plot,

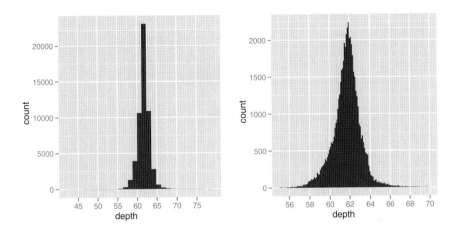

Fig. 5.2: (Left) Never rely on the default parameters to get a revealing view of the distribution. (Right) Zooming in on the x axis, `xlim = c(55, 70)`, and selecting a smaller bin width, `binwidth = 0.1`, reveals far more detail. We can see that the distribution is slightly skew-right. Don't forget to include information about important parameters (like bin width) in the caption.

`position = "fill"`. These options are illustrated in Figure 5.3, created with the code below.

```
depth_dist <- ggplot(diamonds, aes(depth)) + xlim(58, 68)
depth_dist +
  geom_histogram(aes(y = ..density..), binwidth = 0.1) +
  facet_grid(cut ~ .)
depth_dist + geom_histogram(aes(fill = cut), binwidth = 0.1,
  position = "fill")
depth_dist + geom_freqpoly(aes(y = ..density.., colour = cut),
  binwidth = 0.1)
```

Both the histogram and frequency polygon geom use `stat_bin`. This statistic produces two output variables `count` and `density`. The count is the default as it is most interpretable. The density is basically the count divided by the total count, and is useful when you want to compare the shape of the distributions, not the overall size. You will often prefer this when comparing the distribution of subsets that have different sizes.

Many of the distribution-related geoms come in geom/stat pairs. Most of these geoms are aliases: a basic geom is combined with a stat to produce the desired plot. The boxplot may appear to be an exception to this rule, but behind the scenes `geom_boxplot` uses a combination of the basic bars, lines and points.

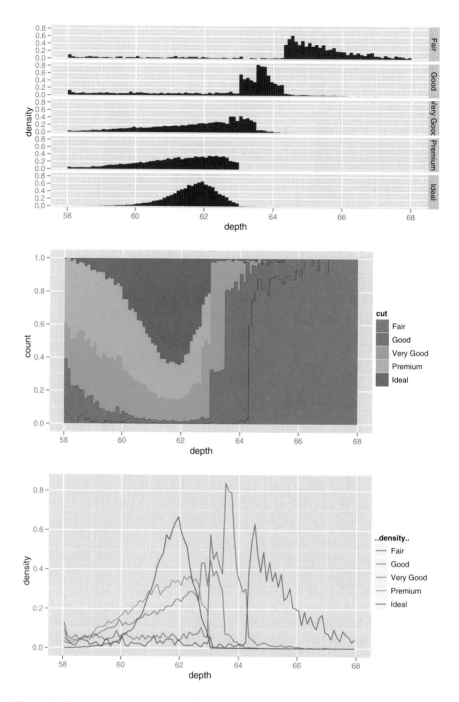

Fig. 5.3: Three views of the distribution of depth and cut. From top to bottom: faceted histogram, a conditional density plot, and frequency polygons. All show an interesting pattern: as quality increases, the distribution shifts to the left and becomes more symmetric.

- `geom_boxplot` = `stat_boxplot` + `geom_boxplot`: box-and-whisker plot, for a continuous variable conditioned by a categorical variable. This is a useful display when the categorical variable has many distinct values. When there are few values, the techniques described above give a better view of the shape of the distribution. This technique can also be used for continuous variables, if they are first finely binned. Figure 5.4 shows boxplots conditioned on both categorical and continuous variables.

```
qplot(cut, depth, data=diamonds, geom="boxplot")
qplot(carat, depth, data=diamonds, geom="boxplot",
    group = round_any(carat, 0.1, floor), xlim = c(0, 3))
```

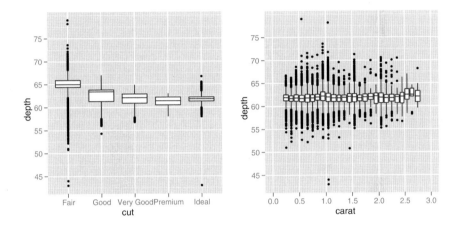

Fig. 5.4: The boxplot geom can be use to see the distribution of a continuous variable conditional on a discrete varable like cut (left), or continuous variable like carat (right). For continuous variables, the group aesthetic must be set to get multiple boxplots. Here `group = round_any(carat, 0.1, floor)` is used to get a boxplot for each 0.1 carat bin.

- `geom_jitter` = `position_jitter` + `geom_point`: a crude way of looking at discrete distributions by adding random noise to the discrete values so that they don't overplot. An example is shown in Figure 5.5 created with the code below.

```
qplot(class, cty, data=mpg, geom="jitter")
qplot(class, drv, data=mpg, geom="jitter")
```

- `geom_density` = `stat_density` + `geom_area`: a smoothed version of the frequency polygon based on kernel smoothers. Also described in Section 2.5.3. Use a density plot when you know that the underlying density is smooth, continuous and unbounded. You can use the `adjust` parameter to

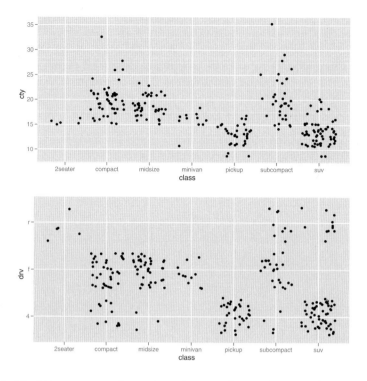

Fig. 5.5: The jitter geom can be used to give a crude visualisation of 2d distributions with a discrete component. Generally this works better for smaller datasets. Car class vs. continuous variable city mpg (top) and discrete variable drive train (bottom).

make the density more or less smooth. An example is shown in Figure 5.6 created with the code below.

```
qplot(depth, data=diamonds, geom="density", xlim = c(54, 70))
qplot(depth, data=diamonds, geom="density", xlim = c(54, 70),
  fill = cut, alpha = I(0.2))
```

Visualising a joint 2d continuous distribution is described in the next section.

5.5 Dealing with overplotting

The scatterplot is a very important tool for assessing the relationship between two continuous variables. However, when the data is large, often points will be plotted on top of each other, obscuring the true relationship. In extreme cases, you will only be able to see the extent of the data, and any conclusions

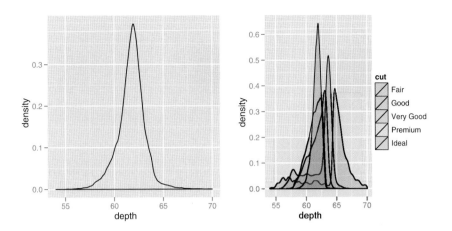

Fig. 5.6: The density plot is a smoothed version of the histogram. It has desirable theoretical properties, but is more difficult to relate back to the data. A density plot of depth (left), coloured by cut (right).

drawn from the graphic will be suspect. This problem is called overplotting and there are a number of ways to deal with it:

- Small amounts of overplotting can sometimes be alleviated by making the points smaller, or using hollow glyphs, as shown in Figure 5.7. The data is 2000 points sampled from two independent normal distributions, and the code to produce the graphic is shown below.

```
df <- data.frame(x = rnorm(2000), y = rnorm(2000))
norm <- ggplot(df, aes(x, y))
norm + geom_point()
norm + geom_point(shape = 1)
norm + geom_point(shape = ".") # Pixel sized
```

- For larger datasets with more overplotting, you can use alpha blending (transparency) to make the points transparent. If you specify alpha as a ratio, the denominator gives the number of points that must be overplotted to give a solid colour. In R, the lowest amount of transparency you can use is 1/256, so it will not be effective for heavy overplotting. Figure 5.8 demonstrates some of these options with the following code.

```
norm + geom_point(colour = alpha("black", 1/3))
norm + geom_point(colour = alpha("black", 1/5))
norm + geom_point(colour = alpha("black", 1/10))
```

- If there is some discreteness in the data, you can randomly jitter the points to alleviate some overlaps. This is particularly useful in conjunction

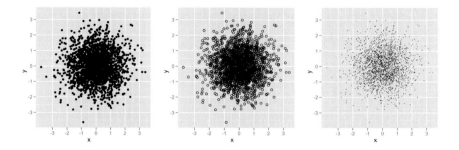

Fig. 5.7: Modifying the glyph used can help with mild to moderate overplotting. From left to right: the default shape, `shape = 1` (hollow points), and `shape= "."` (pixel points).

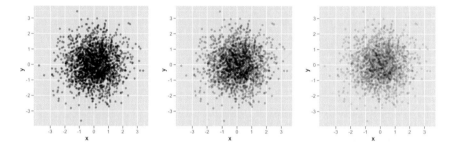

Fig. 5.8: Using alpha blending to alleviate overplotting in sample data from a bivariate normal. Alpha values from left to right: 1/3, 1/5, 1/10.

with transparency. By default, the amount of jitter added is 40% of the resolution of the data, which leaves a small gap between adjacent regions. In Figure 5.9, table is recorded to the nearest integers, so we set a jitter width of half of that. The complete code is shown below.

```
td <- ggplot(diamonds, aes(table, depth)) +
  xlim(50, 70) + ylim(50, 70)
td + geom_point()
td + geom_jitter()
jit <- position_jitter(width = 0.5)
td + geom_jitter(position = jit)
td + geom_jitter(position = jit, colour = alpha("black", 1/10))
td + geom_jitter(position = jit, colour = alpha("black", 1/50))
td + geom_jitter(position = jit, colour = alpha("black", 1/200))
```

Alternatively, we can think of overplotting as a 2d density estimation problem, which gives rise to two more approaches:

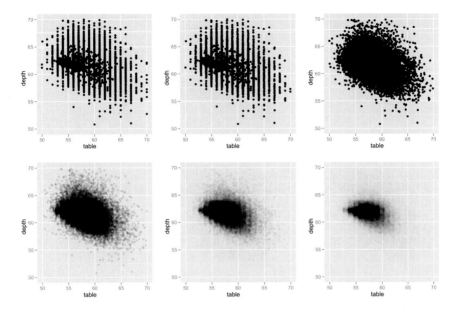

Fig. 5.9: A plot of table vs. depth from the diamonds data, showing the use of jitter and alpha blending to alleviate overplotting in discrete data. From left to right: geom point, geom jitter with default jitter, geom jitter with horizontal jitter of 0.5 (half the gap between bands), alpha of 1/10, alpha of 1/50, alpha of 1/200.

- Bin the points and count the number in each bin, then visualise that count in some way (the 2d generalisation of the histogram). Breaking the plot into many small squares can produce distracting visual artefacts. Carr et al. (1987) suggests using hexagons instead, and this is implemented with `geom_hexagon`, using the capabilities of the `hexbin` package (Carr et al., 2008). Figure 5.10 compares square and hexagonal bins, using parameters `bins` and `binwidth` to control the number and size of the bins. The complete code is shown below.

```
d <- ggplot(diamonds, aes(carat, price)) + xlim(1,3) +
  opts(legend.position = "none")
d + stat_bin2d()
d + stat_bin2d(bins = 10)
d + stat_bin2d(binwidth=c(0.02, 200))
d + stat_binhex()
d + stat_binhex(bins = 10)
d + stat_binhex(binwidth=c(0.02, 200))
```

- Estimate the 2d density with `stat_density2d`, and overlay contours from this distribution on the scatterplot, or display the density by itself as

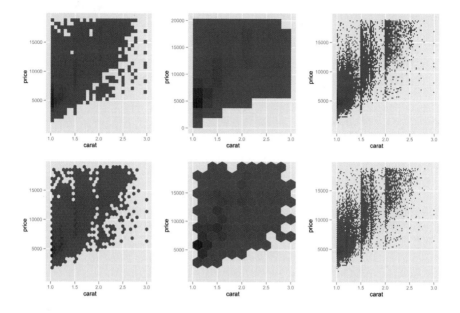

Fig. 5.10: Binning with, top row, square bins, and bottom row, hexagonal bins.
Left column uses default parameters, middle column `bins = 10`, and right column
`binwidth = c(0.02, 200)`. Legends have been omitted to save space.

coloured tiles, or points with size proportional to density. Figure 5.11 shows
a few of these options with the code below.

```
d <- ggplot(diamonds, aes(carat, price)) + xlim(1,3) +
  opts(legend.position = "none")
d + geom_point() + geom_density2d()
d + stat_density2d(geom = "point", aes(size = ..density..),
 contour = F) + scale_area(to = c(0.2, 1.5))
d + stat_density2d(geom = "tile", aes(fill = ..density..),
  contour = F)
last_plot() + scale_fill_gradient(limits = c(1e-5,8e-4))
```

• If you are interested in the conditional distribution of y given x, then the
  techniques of Section 2.5.3 will also be useful.

Another approach to dealing with overplotting is to add data summaries
to help guide the eye to the true shape of the pattern within the data. For
example, you could add a smooth line showing the centre of the data with
`geom_smooth`. Section 5.9 has more ideas.

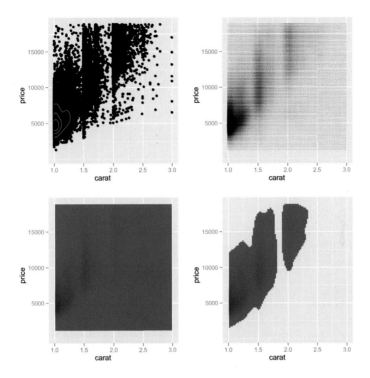

Fig. 5.11: Using density estimation to model and visualise point densities. (Top) Image displays of the density; (bottom) point and contour based displays.

5.6 Surface plots

ggplot2 currently does not support true 3d surfaces. However, it does support the common tools for representing 3d surfaces in 2d: contours, coloured tiles and bubble plots. These were used to illustrated the 2d density surfaces in the previous section. You may also want to look at RGL, http://rgl. neoscientists.org/about.shtml, for interactive 3d plots, including true 3d surfaces.

5.7 Drawing maps

ggplot2 provides some tools to make it easy to combine maps from the maps package with other ggplot2 graphics. Table 5.1 lists the available maps, which are unfortunately rather US centric. There are two basic reasons you might want to use map data: to add reference outlines to a plot of spatial data, or to construct a choropleth map by filling regions with colour.

Adding map border is performed by the borders() function. The first two arguments select the map and region within the map to display. The

Country	Map name
France	france
Italy	italy
New Zealand	nz
USA at county level	county
USA at state level	state
USA borders	usa
Entire world	world

Table 5.1: Maps available in the maps package

remaining arguments control the appearance of the borders: their `colour` and `size`. If you'd prefer filled polygons instead of just borders, you can set the `fill` colour. The following code uses `borders()` to display the spatial data shown in Figure 5.12.

```
library(maps)
data(us.cities)
big_cities <- subset(us.cities, pop > 500000)
qplot(long, lat, data = big_cities) + borders("state", size = 0.5)

tx_cities <- subset(us.cities, country.etc == "TX")
ggplot(tx_cities, aes(long, lat)) +
  borders("county", "texas", colour = "grey70") +
  geom_point(colour = alpha("black", 0.5))
```

Choropleth maps are a little trickier and a lot less automated because it is challenging to match the identifiers in your data to the identifiers in the map data. The following example shows how to use `map_data()` to convert a map into a data frame, which can then be `merge()`d with your data to produce a choropleth map. The results are shown in Figure 5.13. The details for your data will probably be different, but the key is to have a column in your data and a column in the map data that can be matched.

```
library(maps)
states <- map_data("state")
arrests <- USArrests
names(arrests) <- tolower(names(arrests))
arrests$region <- tolower(rownames(USArrests))

choro <- merge(states, arrests, by = "region")
# Reorder the rows because order matters when drawing polygons
# and merge destroys the original ordering
choro <- choro[order(choro$order), ]
qplot(long, lat, data = choro, group = group,
```

Fig. 5.12: Example using the borders function. (Left) All cities with population (as of January 2006) of greater than half a million, (right) cities in Texas.

```
  fill = assault, geom = "polygon")
qplot(long, lat, data = choro, group = group,
  fill = assault / murder, geom = "polygon")
```

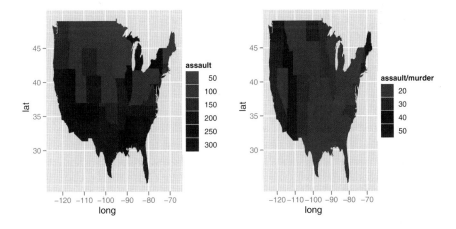

Fig. 5.13: Two choropleth maps showing number of assaults (left) and the ratio of assaults to murders (right).

The `map_data()` function is also useful if you'd like to process the map data in some way. In the following example we compute the (approximate) centre of each county in Iowa and then use those centres to label the map.

```
> ia <- map_data("county", "iowa")
> mid_range <- function(x) mean(range(x, na.rm = TRUE))
> centres <- ddply(ia, .(subregion),
+   colwise(mid_range, .(lat, long)))
> ggplot(ia, aes(long, lat)) +
+   geom_polygon(aes(group = group),
+     fill = NA, colour = "grey60") +
+   geom_text(aes(label = subregion), data = centres,
+     size = 2, angle = 45)
```

5.8 Revealing uncertainty

If you have information about the uncertainty present in your data, whether it be from a model or from distributional assumptions, it is often important to display it. There are four basic families of geoms that can be used for this job, depending on whether the x values are discrete or continuous, and whether or not you want to display the middle of the interval, or just the extent. These geoms are listed in Table 5.2. These geoms assume that you are interested in the distribution of y conditional on x and use the aesthetics ymin and ymax to determine the range of the y values. If you want the opposite, see coord_flip, Section 7.3.3.

X variable	Range	Range plus centre
Continuous	geom_ribbon	geom_smooth(stat="identity")
Discrete	geom_errorbar	geom_crossbar
	geom_linerange	geom_pointrange

Table 5.2: Geoms that display intervals, useful for visualising uncertainty.

Because there are so many different ways to calculate standard errors, the calculation is up to you. For very simple cases, `ggplot2` provides some tools in the form of summary functions described in Section 5.9, otherwise you will have to do it yourself. The `effects` package (Fox, 2008) is particularly useful for extracting these values from linear models. The following example fits a two-way model with interaction, and shows how to extract and visualise marginal and conditional effects. Figure 5.15 focusses on the categorical variable colour, and Figure 5.16 focusses on the continuous variable carat.

```
> d <- subset(diamonds, carat < 2.5 &
+   rbinom(nrow(diamonds), 1, 0.2) == 1)
> d$lcarat <- log10(d$carat)
> d$lprice <- log10(d$price)
>
> # Remove overall linear trend
> detrend <- lm(lprice ~ lcarat, data = d)
> d$lprice2 <- resid(detrend)
>
> mod <- lm(lprice2 ~ lcarat * color, data = d)
>
> library(effects)
> effectdf <- function(...) {
+   suppressWarnings(as.data.frame(effect(...)))
+ }
> color <- effectdf("color", mod)
> both1 <- effectdf("lcarat:color", mod)
>
> carat <- effectdf("lcarat", mod, default.levels = 50)
> both2 <- effectdf("lcarat:color", mod, default.levels = 3)
```

Note, when captioning such figures, you need to carefully describe the nature of the confidence intervals, and whether or not it is meaningful to look at the overlap. That is, are the standard errors for the means or for the differences between means? The packages `multcomp` and `multcompView` are useful calculating and displaying these errors while correctly adjusting for multiple comparisons.

5.9 Statistical summaries

It's often useful to be able to summarise the y values for each unique x value. In `ggplot2`, this role is played by `stat_summary()`, which provides a flexible way of summarising the conditional distribution of y with the aesthetics `ymin`, `y` and `ymax`. Figure 5.17 shows some of the variety of summaries that can be achieved with this tool.

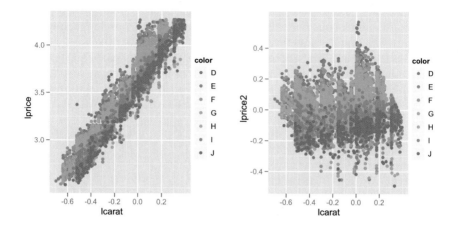

Fig. 5.14: Data transformed to remove most obvious effects. (Left) Both x and y axes are log10 transformed to remove non-linearity. (Right) The major linear trend is removed.

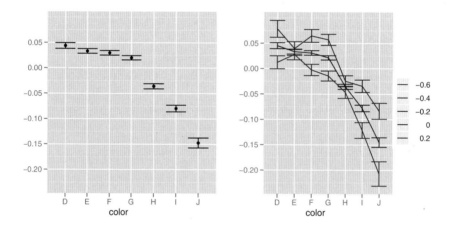

Fig. 5.15: Displaying uncertainty in model estimates for colour. (Left) Marginal effect of colour. (Right) conditional effects of colour for different levels of carat. Error bars show 95% pointwise confidence intervals.

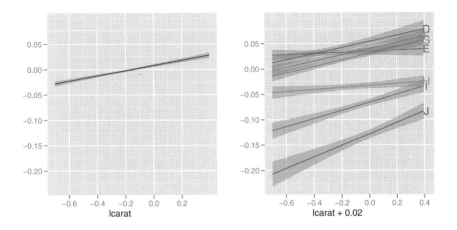

Fig. 5.16: Displaying uncertainty in model estimates for carat. (Left) marginal effect of carat. (Right) conditional effects of carat for different levels of colour. Bands show 95% point-wise confidence intervals.

When using `stat_summary()` you can either supply these the summary functions individually or altogether. These alternatives are described below.

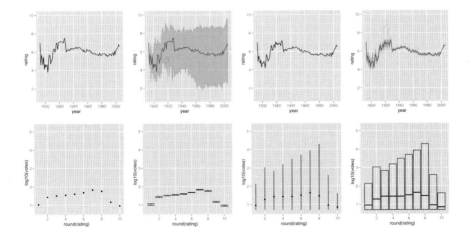

Fig. 5.17: Examples of `stat_summary` in use. (Top) Continuous x with, from left to right, median and line, `median_hilow()` and smooth, mean and line, and `mean_cl_boot()` and smooth. (Bottom) Discrete x with, from left to right, `mean()` and point, `mean_cl_normal()` and error bar, `median_hilow()` and point range, and `median_hilow()` and crossbar. Note that `ggplot2` displays the full range of the data, not just the range of the summary statistics.

5.9.1 Individual summary functions

The arguments `fun.y`, `fun.ymin` and `fun.ymax` accept simple numeric summary functions. You can use any summary function that takes a vector of numbers and returns a single numeric value: `mean()`, `median()`, `min()`, `max()`.

```
> midm <- function(x) mean(x, trim = 0.5)
> m2 +
+   stat_summary(aes(colour = "trimmed"), fun.y = midm,
+     geom = "point") +
+   stat_summary(aes(colour = "raw"), fun.y = mean,
+     geom = "point") +
+   scale_colour_hue("Mean")
```

5.9.2 Single summary function

`fun.data` can be used with more complex summary functions such as one of the summary functions from the `Hmisc` package (Harrell, 2008) described in Table 5.3. You can also write your own summary function. This summary function should return a named vector as output, as shown in the following example.

```
> iqr <- function(x, ...) {
+   qs <- quantile(as.numeric(x), c(0.25, 0.75), na.rm = T)
+   names(qs) <- c("ymin", "ymax")
+   qs
+ }
> m + stat_summary(fun.data = "iqr", geom="ribbon")
```

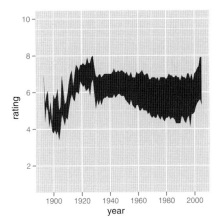

Function	Hmisc original	Middle	Range
mean_cl_normal()	smean.cl.boot()	Mean	Standard error from normal approximation
mean_cl_boot()	smean.cl.boot()	Mean	Standard error from bootstrap
mean_sdl()	smean.sdl()	Mean	Multiple of standard deviation
median_hilow()	smedian.hilow()	Median	Outer quantiles with equal tail areas

Table 5.3: Summary functions from the Hmisc package that have special wrappers to make them easy to use with stat_summary().

5.10 Annotating a plot

When annotating your plot with additional labels, the important thing to remember is that these annotations are just extra data. There are two basic ways to add annotations: one at a time, or many at once.

Adding one at a time works best for small numbers of annotations with varying aesthetics. You just set all the values to give the desired properties. If you have multiple annotations with similar properties, it may make sense to put them all in a data frame and add them at once. The example below demonstrates both approaches by adding information about presidents to economic data.

```
> (unemp <- qplot(date, unemploy, data=economics, geom="line",
+   xlab = "", ylab = "No. unemployed (1000s)"))
```

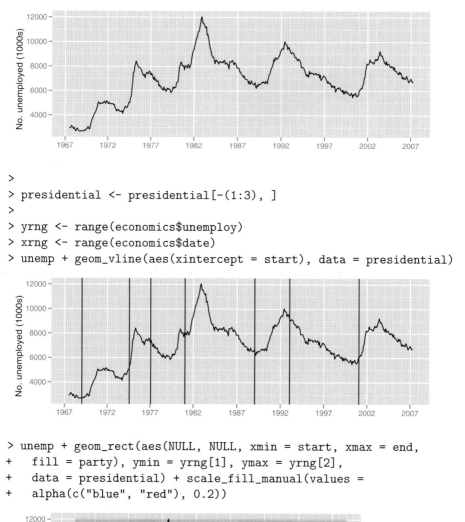

```
>
> presidential <- presidential[-(1:3), ]
>
> yrng <- range(economics$unemploy)
> xrng <- range(economics$date)
> unemp + geom_vline(aes(xintercept = start), data = presidential)
```

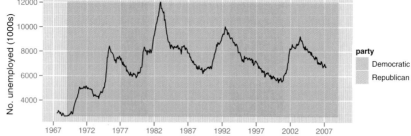

```
> unemp + geom_rect(aes(NULL, NULL, xmin = start, xmax = end,
+    fill = party), ymin = yrng[1], ymax = yrng[2],
+    data = presidential) + scale_fill_manual(values =
+    alpha(c("blue", "red"), 0.2))
```

```
> last_plot() + geom_text(aes(x = start, y = yrng[1], label = name),
+    data = presidential, size = 3, hjust = 0, vjust = 0)
```

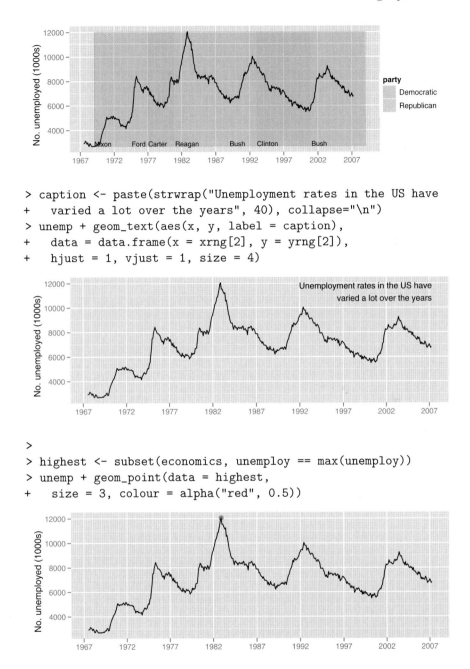

```
> caption <- paste(strwrap("Unemployment rates in the US have
+   varied a lot over the years", 40), collapse="\n")
> unemp + geom_text(aes(x, y, label = caption),
+   data = data.frame(x = xrng[2], y = yrng[2]),
+   hjust = 1, vjust = 1, size = 4)
```

```
>
> highest <- subset(economics, unemploy == max(unemploy))
> unemp + geom_point(data = highest,
+   size = 3, colour = alpha("red", 0.5))
```

- geom_text for adding text descriptions or labelling points. Most plots will not benefit from adding text to every single observation on the plot.

However, pulling out just a few observations (using subset) can be very useful. Typically you will want to label outliers or other important points.

- `geom_vline`, `geom_hline`: add vertical or horizontal lines to a plot.
- `geom_abline`: add lines with arbitrary slope and intercept to a plot.
- `geom_rect` for highlighting interesting rectangular regions of the plot. `geom_rect` has aesthetics `xmin`, `xmax`, `ymin` and `ymax`.
- `geom_line`, `geom_path` and `geom_segment` for adding lines. All these geoms have an `arrow` parameter, which allows you to place an arrowhead on the line. You create arrowheads with the `arrow()` function, which has arguments `angle`, `length`, `ends` and `type`.

5.11 Weighted data

When you have aggregated data where each row in the dataset represents multiple observations, you need some way to take into account the weighting variable. We will use some data collected on Midwest states in the 2000 US census. The data consists mainly of percentages (e.g., percent white, percent below poverty line, percent with college degree) and some information for each county (area, total population, population density).

There are a few different things we might want to weight by:

- nothing, to look at numbers of counties
- total population, to work with absolute numbers
- area, to investigate geographic effects

The choice of a weighting variable profoundly affects what we are looking at in the plot and the conclusions that we will draw. There are two aesthetic attributes that can be used to adjust for weights. Firstly, for simple geoms like lines and points, you can make the size of the grob proportional to the number of points, using the `size` aesthetic, as with the following code, whose results are shown in Figure 5.18.

```
qplot(percwhite, percbelowpoverty, data = midwest)
qplot(percwhite, percbelowpoverty, data = midwest,
  size = poptotal / 1e6) + scale_area("Population\n(millions)",
  breaks = c(0.5, 1, 2, 4))
qplot(percwhite, percbelowpoverty, data = midwest, size = area) +
  scale_area()
```

For more complicated grobs which involve some statistical transformation, we specify weights with the `weight` aesthetic. These weights will be passed on to the statistical summary function. Weights are supported for every case where it makes sense: smoothers, quantile regressions, boxplots, histograms, and density plots. You can't see this weighting variable directly, and it doesn't produce a legend, but it will change the results of the statistical summary.

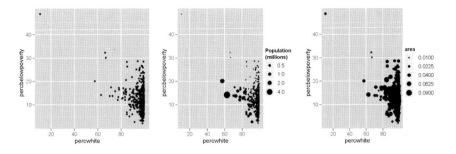

Fig. 5.18: Using size to display weights. No weighting (left), weighting by population (centre) and by area (right).

Figure 5.19 shows how weighting by population density affects the relationship between percent white and percent below the poverty line.

```
lm_smooth <- geom_smooth(method = lm, size = 1)
qplot(percwhite, percbelowpoverty, data = midwest) + lm_smooth
qplot(percwhite, percbelowpoverty, data = midwest,
  weight = popdensity, size = popdensity) + lm_smooth
```

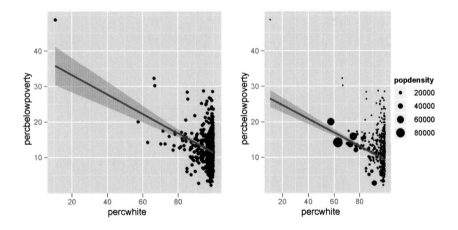

Fig. 5.19: An unweighted line of best fit (left) and weighted by population size (right).

When we weight a histogram or density plot by total population, we change from looking at the distribution of the number of counties, to the distribution of the number of people. Figure 5.20 shows the difference this makes for a histogram of the percentage below the poverty line.

```
qplot(percbelowpoverty, data = midwest, binwidth = 1)
```

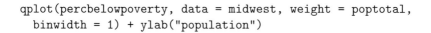

```
qplot(percbelowpoverty, data = midwest, weight = poptotal,
  binwidth = 1) + ylab("population")
```

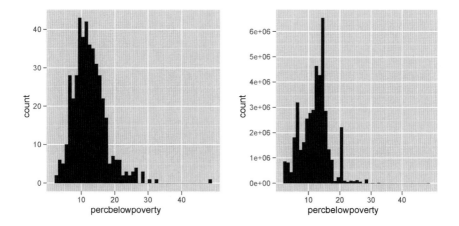

Fig. 5.20: The difference between an unweighted (left) and weighted (right) histogram. The unweighted histogram shows number of counties, while the weighted histogram shows population. The weighting considerably changes the interpretation!

Chapter 6

Scales, axes and legends

6.1 Introduction

Scales control the mapping from data to aesthetics. They take your data and turn it into something that you can perceive visually: e.g., size, colour, position or shape. Scales also provide the tools you use to read the plot: the axes and legends (collectively known as guides). Formally, each scale is a function from a region in data space (the domain of the scale) to a region in aesthetic space (the range of the range). The domain of each scale corresponds to the range of the variable supplied to the scale, and can be continuous or discrete, ordered or unordered. The range consists of the concrete aesthetics that you can perceive and that R can understand: position, colour, shape, size and line type. If you blinked when you read that scales map data both to position and colour, you are not alone. The notion that the same kind of object is used to map data to positions and symbols strikes some people as unintuitive. However, you will see the logic and power of this notion as you read further in the chapter.

The process of scaling takes place in three steps, transformation, training and mapping, and is described in Section 6.2. Without a scale, there is no way to go from the data to aesthetics, so a scale is required for every aesthetic used on the plot. It would be tedious to manually add a scale every time you used a new aesthetic, so whenever a scale is needed ggplot2 will add a default. You can generate many plots without knowing how scales work, but understanding scales and learning how to manipulate them will give you much more control. Default scales and how to override them are described in Section 6.3.

Scales can be roughly divided into four categories: position scales, colour scales, the manual discrete scale and the identity scale. The common options and most important uses are described in Section 6.4. The section focusses on giving you a high-level overview of the options available, rather than expanding on every detail in depth. Details about individual parameters are included in the online documentation.

The other important role of each scale is to produce a **guide** that allows the viewer to perform the inverse mapping, from aesthetic space to data space,

H. Wickham, *ggplot2*, Use R, DOI 10.1007/978-0-387-98141-3_6,

and read values off the plot. For position aesthetics, the axes are the guides; for all other aesthetics, legends do the job. Unlike other plotting systems, you have little direct control over the axis or legend: there is no `gglegend()` or `ggaxis()` to call to modify legends or axes. Instead, all aspects of the guides are controlled by parameters of the scale. Axes and legends are discussed in Section 6.5.

Section 6.6 concludes the chapter with pointers to other academic work that discusses some of the things you need to keep in mind when assigning variables to aesthetics.

6.2 How scales work

To describe how scales work, we will first describe the domain (the data space) and the range (the aesthetic space), and then outline the process by which one is mapped to the other.

Since an input variable is either discrete or continuous, the domain is either a set of values (stored as a factor, character vector or logical vector) or an interval on the real line (stored as a numeric vector of length 2). For example, in the mammals sleep dataset (`msleep`), the domain of the discrete variable vore is {carni, herbi, omni}, and the domain of the continuous variable bodywt is [0.005, 6654]. We often think of these as data ranges, but here we are focussing on their nature as input to the scale, i.e., as a domain of a function.

The range can also be discrete or continuous. For discrete scales, it is a vector of aesthetic values corresponding to the input values. For continuous scales, it is a 1d path through some more complicated space. For example, a colour gradient interpolates linearly from one colour to another. The range is either specified by the user when the scale is created, or by the scale itself.

The process of mapping the domain to the range includes the following stages:

- **transformation**: (for continuous domain only). It is often useful to display a transformation of the data, such as a logarithm or square root. Transformations are described in more depth in Section 6.4.2.
  After any transformations have been applied, the statistical summaries for each layer are computed based on the transformed data. This ensures that a plot of $\log(x)$ vs. $\log(y)$ on linear scales looks the same as x vs. y on log scales.
- **training**: During this key stage, the domain of the scale is learned. Sometimes learning the domain of a scale is extremely straightforward: In a plot with only one layer, representing only raw data, it consists of determining the minimum and maximum values of a continuous variable (after transformation), or listing the unique levels of a categorical variable. However, often the domain must reflect multiple layers across multiple datasets in multiple panels. For example, imagine a scale that will be used to create an

axis; the minimum and maximum values of the raw data in the first layer and the statistical summary in the second layer are likely to be different, but they must all eventually be drawn on the same plot.

The domain can also be specified directly, overriding the training process, by manually setting the domain of the scale with the `limits` argument, as described in Section 6.3. Any values outside of the domain of the scale will be mapped to `NA`.

- **mapping:** We now know the domain and we already knew the range before we started this process, so the last thing to do is to apply the scaling function that maps data values to aesthetic values.

We have left a few stages out of this description of the process for simplicity. For example, we haven't discussed the role faceting plays in training, and we have also ignored position adjustments. Nevertheless this description is accurate, and you should come back to it if you are confused about what scales are doing in your plot.

6.3 Usage

Every aesthetic has a default scale that is added to the plot whenever you use that aesthetic. These are listed in Table 6.1. The scale depends on the variable type: continuous (numeric) or discrete (factor, logical, character). If you want to change the default scales see `set_default_scale()`, described in Section 8.2.1.

Default scales are added when you initialise the plot and when you add new layers. This means it is possible to get a mismatch between the variable type and the scale type if you later modify the underlying data or aesthetic mappings. When this happens you need to add the correct scale yourself. The following example illustrates the problem and solution.

```
plot <- qplot(cty, hwy, data = mpg)
plot

# This doesn't work because there is a mismatch between the
# variable type and the default scale
plot + aes(x = drv)

# Correcting the default manually resolves the problem.
plot + aes(x = drv) + scale_x_discrete()
```

To add a different scale or to modify some features of the default scale, you must construct a new scale and then add it to the plot using `+`. All scale constructors have a common naming scheme. They start with `scale_`, followed by the name of the aesthetic (e.g., `colour_`, `shape_` or `x_`), and finally by the name of the scale (e.g., `gradient`, `hue` or `manual`). For example, the

name of the default scale for the colour aesthetic based on discrete data is
scale_colour_hue(), and the name of the Brewer colour scale for fill colour
is scale_fill_brewer().

Aesthetic	Discrete	Continuous
Colour and fill	brewer	**gradient**
	grey	gradient2
	hue	gradientn
	identity	
	manual	
Position (x, y)	**discrete**	**continuous**
		date
Shape	**shape**	
	identity	
	manual	
Line type	**linetype**	
	identity	
	manual	
Size	identity	**size**
	manual	

Table 6.1: Scales, by aesthetic and variable type. Default scales are emboldened.
The default scale varies depending on whether the variable is continuous or discrete.
Shape and line type do not have a default continuous scale; size does not have a
default discrete scale.

The following code illustrates this process. We start with a plot that uses the
default colour scale, and then modify it to adjust the appearance of the legend,
and then use a different colour scale. The results are shown in Figure 6.1.

```
p <- qplot(sleep_total, sleep_cycle, data = msleep, colour = vore)
p
# Explicitly add the default scale
p + scale_colour_hue()

# Adjust parameters of the default, here changing the appearance
# of the legend
p + scale_colour_hue("What does\nit eat?",
    breaks = c("herbi", "carni", "omni", NA),
    labels = c("plants", "meat", "both", "don't know"))

# Use a different scale
p + scale_colour_brewer(pal = "Set1")
```

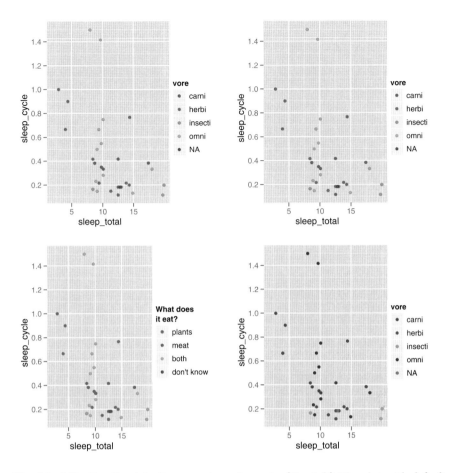

Fig. 6.1: Adjusting the default parameters of a scale. (Top left) The plot with default scale. (Top right) Adding the default scale by hand doesn't change the appearance of the plot. (Bottom left) Adjusting the parameters of the scale to tweak the legend. (Bottom right) Using a different colour scale: Set1 from the ColorBrewer colours.

6.4 Scale details

Scales can be divided roughly into four separate groups:

- Position scales, used to map continuous, discrete and date-time variables onto the plotting region and to construct the corresponding axes.
- Colour scales, used to map continuous and discrete variables to colours.
- Manual scales, used to map discrete variables to your choice of symbol size, line type, shape or colour, and to create the corresponding legend.
- The identity scale, used to plot variable values directly to the aesthetic rather than mapping them. For example, if the variable you want to map to symbol colour is itself a vector of colours, you want to render those values directly rather than mapping them to some other colours.

This section describes each group in more detail. Precise details about individual scales can be found in the documentation, within R (e.g., `?scale_brewer`), or online at `http://had.co.nz/ggplot2`.

6.4.1 Common arguments

The following arguments are common to all scales.

- `name`: sets the label which will appear on the axis or legend. You can supply text strings (using `\n` for line breaks) or mathematical expressions (as described by `?plotmath`). Because tweaking these labels is such a common task, there are three helper functions to save you some typing: `xlab()`, `ylab()` and `labs()`. Their use is demonstrated in the code below and results are shown in Figure 6.2.

```
p <- qplot(cty, hwy, data = mpg, colour = displ)
p
p + scale_x_continuous("City mpg")
p + xlab("City mpg")
p + ylab("Highway mpg")
p + labs(x = "City mpg", y = "Highway", colour = "Displacement")
p + xlab(expression(frac(miles, gallon)))
```

- `limits`: fixes the domain of the scale. Continuous scales take a numeric vector of length two; discrete scales take a character vector. If limits are set, no training of the data will be performed. See Section 6.4.2 for shortcuts. Limits are useful for removing data you don't want displayed in a plot (i.e., setting limits that are smaller than the full range of data), and for ensuring that limits are consistent across multiple plots intended to be compared (i.e., setting limits that are larger or smaller than some of the default ranges).

  Any value not in the domain of the scale is discarded: for an observation to be included in the plot, each aesthetic must be in the domain of each scale. This discarding occurs before statistics are calculated.

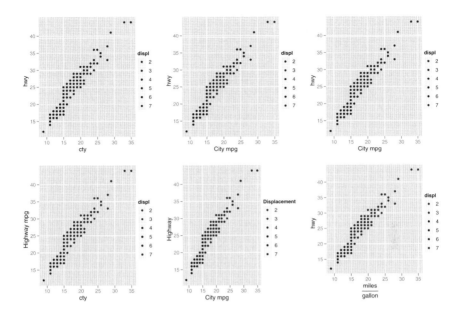

Fig. 6.2: A demonstration of the different forms legend title can take.

- breaks and labels: breaks controls which values appear on the axis or legend, i.e., what values tick marks should appear on an axis or how a continuous scale is segmented in a legend. labels specifies the labels that should appear at the breakpoints. If labels is set, you must also specify breaks, so that the two can be matched up correctly.

  To distinguish breaks from limits, remember that breaks affect what appears on the axes and legends, while limits affect what appears on the plot. See Figure 6.3 for an illustration. The first column uses the default settings for both breaks and limits, which are limits = c(4, 8) and breaks = 4:8. In the middle column, the breaks have been reset: the plotted region is the same, but the tick positions and labels have shifted. In the right column, it is the limits which have been redefined, so much of the data now falls outside the plotting region.

```
p <- qplot(cyl, wt, data = mtcars)
p
p + scale_x_continuous(breaks = c(5.5, 6.5))
p + scale_x_continuous(limits = c(5.5, 6.5))
p <- qplot(wt, cyl, data = mtcars, colour = cyl)
p
p + scale_colour_gradient(breaks = c(5.5, 6.5))
p + scale_colour_gradient(limits = c(5.5, 6.5))
```

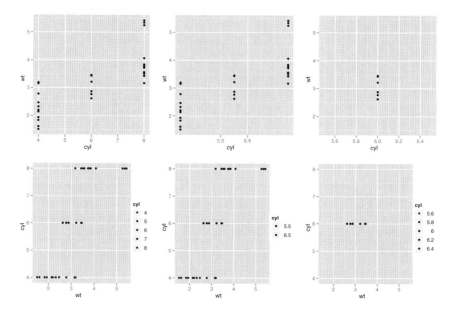

Fig. 6.3: The difference between breaks and limits. (Left) default plot with `limits` = c(4, 8), breaks = 4:8, (middle) breaks = c(5.5,6.5) and (right) limits = c(5.5,6.5). The effect on the x axis (top) and colour legend (bottom)

- **formatter**: if no labels are specified the formatter will be called on each break to produce the label. Useful formatters for continuous scales are `comma`, `percent`, `dollar` and `scientific`, and for discrete scales is `abbreviate`.

6.4.2 Position scales

Every plot must have two position scales, one for the horizontal position (the x scale) and one for the vertical position (the y scale). ggplot2 comes with continuous, discrete (for factor, character and logical vectors) and date scales. Each of these transform the data in a slightly different way, and generate a slightly different type of axis. The following sections describe each type in more detail.

A common task for all position axes is changing the axis limits. Because this is such a common task, ggplot2 provides a couple of helper functions to save you some typing: `xlim()` and `ylim()`. These functions inspect their input and then create the appropriate scale, as follows:

- `xlim(10, 20)`: a continuous scale from 10 to 20
- `ylim(20, 10)`: a reversed continuous scale from 20 to 10
- `xlim("a", "b", "c")`: a discrete scale

- `xlim(as.Date(c("2008-05-01", "2008-08-01")))`: a date scale from May 1 to August 1 2008.

These limits do not work in the same way as `xlim` and `ylim` in base or lattice graphics. In `ggplot2`, to be consistent with the other scales, any data outside the limits is not plotted and not included in the statistical transformation. This means that setting the limits is not the same as visually zooming in to a region of the plot. To do that, you need to use the `xlim` and `ylim` arguments to `coord_cartesian()`, described in Section 7.3.3. This performs purely visual zooming and does not affect the underlying data.

By default, the limits of position scales extend a little past the range of the data. This ensures that the data does not overlap the axes. You can control the amount of expansion with the `expand` argument. This parameter should be a numeric vector of length two. The first element gives the multiplicative expansion, and the second the additive expansion. If you don't want any extra space, use `expand = c(0, 0)`.

Continuous

The most common continuous position scales are `scale_x_continuous` and `scale_y_continuous`, which map data to the x and y axis. The most interesting variations are produced using transformations. Every continuous scale takes a `trans` argument, allowing the specification of a variety of transformations, both linear and non-linear. The transformation is carried out by a "transformer," which describes the transformation, its inverse, and how to draw the labels. Table 6.2 lists some of the more common transformers.

Name	Function $f(x)$	Inverse $f^{-1}(y)$
asn	$\tanh^{-1}(x)$	$\tanh(y)$
exp	e^x	$\log(y)$
identity	x	y
log	$\log(x)$	e^y
log10	$\log_{10}(x)$	10^y
log2	$\log_2(x)$	2^y
logit	$\log(\frac{x}{1-x})$	$\frac{1}{1+e(y)}$
pow10	10^x	$\log_{10}(y)$
probit	$\Phi(x)$	$\Phi^{-1}(y)$
recip	x^{-1}	y^{-1}
reverse	$-x$	$-y$
sqrt	$x^{1/2}$	y^2

Table 6.2: List of built-in transformers.

Transformations are most often used to modify position scales, so there are shortcuts for x, y and z scales: `scale_x_log10()` is equivalent to `scale_x_continuous(trans = "log10")`. The `trans` argument works for any continuous scale, including the colour gradients described below, but the shortcuts only exist for position scales.

Of course, you can also perform the transformation yourself. For example, instead of using `scale_x_log()`, you could plot `log10(x)`. That produces an identical result inside the plotting region, but the the axis and tick labels won't be the same. If you use a transformed scale, the axes will be labelled in the original data space. In both cases, the transformation occurs before the statistical summary. Figure 6.4 illustrates this difference with the following code.

```
qplot(log10(carat), log10(price), data = diamonds)
qplot(carat, price, data = diamonds) +
  scale_x_log10() + scale_y_log10()
```

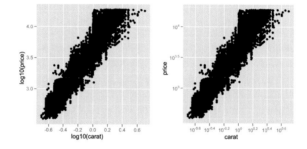

Fig. 6.4: A scatterplot of diamond price vs. carat illustrating the difference between log transforming the scale (left) and log transforming the data (right). The plots are identical, but the axis labels are different.

Transformers are also used in `coord_trans()`, where the transformation occurs after the statistic has been calculated, and affects the shape of the graphical object drawn on the plot. `coord_trans()` is described in more detail in Section 7.3.3.

Date and time

Dates and times are basically continuous values, but with special ways of labelling the axes. Currently, only dates of class `date` and times of class `POSIXct` are supported. If your dates are in a different format you will need to convert them with `as.Date()` or `as.POSIXct()`.

There are three arguments that control the appearance and location of the ticks for date axes: `major`, `minor` and `format`. Generally, the scale does a pretty good job of choosing the defaults, but if you need to tweak them the details are as follows:

- The `major` and `minor` arguments specify the position of major and minor breaks in terms of date units, years, months, weeks, days, hours, minutes and seconds, and can be combined with a multiplier. For example, `major = "2 weeks"` will place a major tick mark every two weeks. If not specified, the date scale has some reasonable default for choosing them automatically.
- The `format` argument specifies how the tick labels should be formatted. Table 6.3 lists the special characters used to display components of a date. For example, if you wanted to display dates of the form 14/10/1979, you would use the string `"%d/%m/%y"`.

Code	Meaning
%S	second (00-59)
%M	minute (00-59)
%l	hour, in 12-hour clock (1-12)
%I	hour, in 12-hour clock (01-12)
%H	hour, in 24-hour clock (00-23)
%a	day of the week, abbreviated (Mon-Sun)
%A	day of the week, full (Monday-Sunday)
%e	day of the month (1-31)
%d	day of the month (01-31)
%m	month, numeric (01-12)
%b	month, abbreviated (Jan-Dec)
%B	month, full (January-December)
%y	year, without century (00-99)
%Y	year, with century (0000-9999)

Table 6.3: Common data formatting codes, adapted from the documentation of `strptime`. Listed from shortest to longest duration.

The code below generates the plots in Figure 6.5, illustrating some of these parameters.

```
plot <- qplot(date, psavert, data = economics, geom = "line") +
  ylab("Personal savings rate") +
  geom_hline(xintercept = 0, colour = "grey50")
plot
plot + scale_x_date(major = "10 years")
plot + scale_x_date(
  limits = as.Date(c("2004-01-01", "2005-01-01")),
  format = "%Y-%m-%d"
)
```

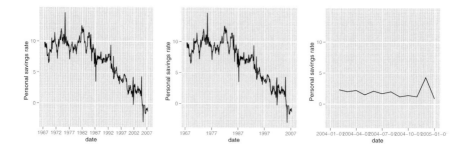

Fig. 6.5: A time series of personal savings rate. (Left) The default appearance, (middle) breaks every 10 years, and (right) scale restricted to 2004, with YMD date format. Measurements are recorded at the end of each month.

Discrete

Discrete position scales map the unique values of their input to integers. The order of the result can be controlled by the `breaks` argument, and levels can be dropped with the `limits` argument (or by using `xlim()` or `ylim()`). Because it is often useful to place labels and other annotations on intermediate positions on the plot, discrete position scales also accept continuous values. If you have not adjusted the breaks or limits, the numerical position of a factor level can be calculated with `as.numeric()`: the values are placed on integers starting at 1.

6.4.3 Colour

After position, probably the most commonly used aesthetic is colour. There are quite a few different ways of mapping values to colours: three different gradient based methods for continuous values, and two methods for mapping discrete values. But before we look at the details of the different methods, it's useful to learn a little bit of colour theory. Colour theory is complex because the underlying biology of the eye and brain is complex, and this introduction will only touch on some of the more important issues. An excellent more detailed exposition is available online at `http://tinyurl.com/clrdtls`.

At the physical level, colour is produced by a mixture of wavelengths of lights. To know a colour completely we need to know the complete mixture of wavelengths, but fortunately for us the human eye only has three different colour receptors, and so we can summarise any colour with just three numbers. You may be familiar with the rgb encoding of colour space, which defines a colour by the intensities of red, green and blue light needed to produce it. One problem with this space is that it is not perceptually uniform: the two colours that are one unit apart may look similar or very different depending on where in the colour space they. This makes it difficult to create a mapping from a continuous variable to a set of colours. There have been many attempts to come up with colours spaces that are more perceptually uniform. We'll use a modern attempt called the hcl colour space, which has three components of **h**ue, **c**hroma and **l**uminance:

- Hue is a number between 0 and 360 (an angle) which gives the "colour" of the colour: like blue, red, orange, etc.
- Luminance is the lightness of the colour. A luminance of 0 produces black, and a luminance of 1 produces white.
- Chroma is the purity of a colour. A chroma of 0 is grey, and the maximum value of chroma varies with luminance.

The combination of these three components does not produce a simple geometric shape. Figure 6.6 attempts to show the 3d shape of the space. Each slice is a constant luminance (brightness) with hue mapped to angle and chroma to radius. You can see the centre of each slice is grey and the colours get more intense as they get closer to the edge.

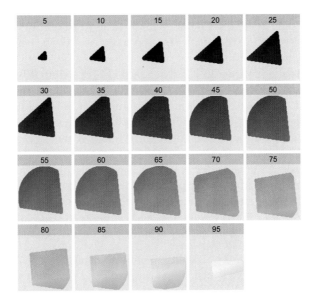

Fig. 6.6: The shape of the hcl colour space. Hue is mapped to angle, chroma to radius and each slice shows a different luminance. The hcl space is a pretty odd shape, but you can see that colours near the centre of each slice are grey, and as you move towards the edges they become more intense. Slices for luminance 0 and 100 are omitted because they would, respectively, be a single black point and a single white point.

An additional complication is that many people ($\sim$10% of men) do not possess the normal complement of colour receptors and so can distinguish fewer colours than usual. In brief, it's best to avoid red-green contrasts, and to check your plots with systems that simulate colour blindness. Visicheck is one online solution. Another alternative is the dichromat package (Lumley, 2007) which provides tools for simulating colour blindness, and a set of colour schemes known to work well for colour-blind people. You can also help people with colour blindness in the same way that you can help people with black-and-white printers: by providing redundant mappings to other aesthetics like size, line type or shape.

All of the scales discussed in the following sections work with border (`colour`) and fill (`fill`) colour aesthetics.

Continuous

There are three types of continuous colour gradients, based on the number of colours in the gradient:

- `scale_colour_gradient()` and `scale_fill_gradient()`: a two-colour gradient, low-high. Arguments `low` and `high` control the colours at either end of the gradient.
- `scale_colour_gradient2()` and `scale_fill_gradient2()`: a three-colour gradient, low-med-high. As well as `low` and `high` colours, these scales also have a `mid` colour for the colour of the midpoint. The midpoint defaults to 0, but can be set to any value with the `midpoint` argument. This is particularly useful for creating diverging colour schemes.
- `scale_colour_gradientn()` and `scale_fill_gradientn()`: a custom n-colour gradient. This scale requires a vector of colours in the `colours` argument. Without further arguments these colours will be evenly spaced along the range of the data. If you want the values to be unequally spaced, use the `values` argument, which should be between 0 and 1 if `rescale` is true (the default), or within the range of the data is `rescale` is false.

Colour gradients are often used to show the height of a 2d surface. In the following example we'll use the surface of a 2d density estimate of the `faithful` dataset (Azzalini and Bowman, 1990), which records the waiting time between eruptions and during each eruption for the Old Faithful geyser in Yellowstone Park. Figure 6.7 shows three gradients applied to this data, created with the following code. Note the use of limits: this parameter is common to all scales.

```
f2d <- with(faithful, MASS::kde2d(eruptions, waiting,
  h = c(1, 10), n = 50))
df <- with(f2d, cbind(expand.grid(x, y), as.vector(z)))
names(df) <- c("eruptions", "waiting", "density")
erupt <- ggplot(df, aes(waiting, eruptions, fill = density)) +
```

```
geom_tile() +
scale_x_continuous(expand = c(0, 0)) +
scale_y_continuous(expand = c(0, 0))
erupt + scale_fill_gradient(limits = c(0, 0.04))
erupt + scale_fill_gradient(limits = c(0, 0.04),
  low = "white", high = "black")
erupt + scale_fill_gradient2(limits = c(-0.04, 0.04),
  midpoint = mean(df$density))
```

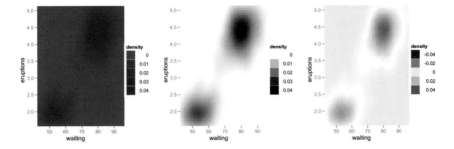

Fig. 6.7: Density of eruptions with three colour schemes. (Left) Default gradient colour scheme, (middle) customised gradient from white to black and (right) 3 point gradient with midpoint set to the mean density.

To create your own custom gradient, use `scale_colour_gradientn()`. This is useful if you have colours that are meaningful for your data (e.g., black body colours or standard terrain colours), or you'd like to use a palette produced by another package. The following code and Figure 6.8 shows palettes generated from routines in the `vcd` package. Zeileis et al. (2008) describes the philosophy behind these palettes and provides a good introduction to some of the complexities of creating good colour scales.

```
library(vcd)
fill_gradn <- function(pal) {
  scale_fill_gradientn(colours = pal(7), limits = c(0, 0.04))
}
erupt + fill_gradn(rainbow_hcl)
erupt + fill_gradn(diverge_hcl)
erupt + fill_gradn(heat_hcl)
```

Discrete

There are two colour scales for discrete data, one which chooses colours in an automated way, and another which makes it easy to select from hand-picked sets.

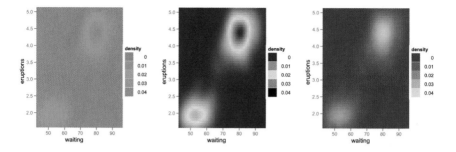

Fig. 6.8: Gradient colour scales using perceptually well-formed palettes produced by the `vcd` package. From left to right: sequential, diverging and heat hcl palettes. Each scale is produced with `scale_fill_gradientn` with colours set to `rainbow_hcl(7)`, `diverge_hcl(7)` and `heat_hcl(7)`.

The default colour scheme, `scale_colour_hue()`, picks evenly spaced hues around the hcl colour wheel. This works well for up to about eight colours, but after that it becomes hard to tell the different colours apart. Another disadvantage of the default colour scheme is that because the colours all have the same luminance and chroma, when you print them in black and white, they all appear as an identical shade of grey.

An alternative to this algorithmic scheme is to use the ColorBrewer colours, http://colorbrewer.org. These colours have been hand picked to work well in a wide variety of situations, although the focus is on maps and so the colours tend to work better when displayed in large areas. For categorical data, the palettes most of interest are "Set1" and "Dark2" for points and "Set2", "Pastel1", "Pastel2" and "Accent" for areas. Use `RColorBrewer::display.brewer.all` to list all palettes. Figure 6.9 shows three of these palettes applied to points and bars, created with the following code.

```
point <- qplot(brainwt, bodywt, data = msleep, log = "xy",
  colour = vore)
area <- qplot(log10(brainwt), data = msleep, fill = vore,
  binwidth = 1)

point + scale_colour_brewer(pal = "Set1")
point + scale_colour_brewer(pal = "Set2")
point + scale_colour_brewer(pal = "Pastel1")
area + scale_fill_brewer(pal = "Set1")
area + scale_fill_brewer(pal = "Set2")
area + scale_fill_brewer(pal = "Pastel1")
```

If you have your own discrete colour scale, you can use `scale_colour_manual()`, as described below.

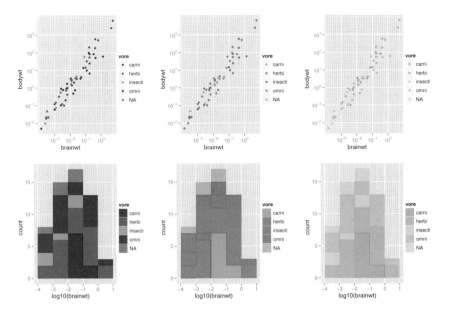

Fig. 6.9: Three ColorBrewer palettes, Set1 (left), Set2 (middle) and Pastel1 (right), applied to points (top) and bars (bottom). Bright colours work well for points, but are overwhelming on bars. Subtle colours work well for bars, but are hard to see on points.

6.4.4 The manual discrete scale

The discrete scales, `scale_linetype()`, `scale_shape()` and `scale_size_discrete` basically have no options (although for the shape scale you can choose whether points should be filled or solid). These scales are just a list of valid values that are mapped to each factor level in turn.

If you want to customise these scales, you need to create your own new scale with the manual scale: `scale_shape_manual()`, `scale_linetype_manual()`, `scale_colour_manual()`, etc. The manual scale has one important argument, `values`, where you specify the values that the scale should produce. If this vector is named, it will match the values of the output to the values of the input, otherwise it will match in order of the levels of the discrete variable. You will need some knowledge of the valid aesthetic values, which are described in Appendix B. The following code demonstrates the use of `scale_manual()`, with results shown in Figure 6.10

```
plot <- qplot(brainwt, bodywt, data = msleep, log = "xy")
plot + aes(colour = vore) +
  scale_colour_manual(value = c("red", "orange", "yellow",
    "green", "blue"))
colours <- c(carni = "red", "NA" = "orange", insecti = "yellow",
```

```
    herbi = "green", omni = "blue")
plot + aes(colour = vore) + scale_colour_manual(value = colours)
plot + aes(shape = vore) +
    scale_shape_manual(value = c(1, 2, 6, 0, 23))
```

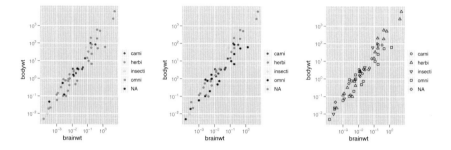

Fig. 6.10: Scale manual used to create custom colour (left and middle) and shape (right) scales.

The following example shows a creative use scale_colour_manual(), when you want to display multiple variables on the same plot, and show a useful legend. In most other plotting systems, you'd just colour the lines as below, and then add a legend that describes which colour corresponds to which variable. That doesn't work in ggplot2 because it's the scales that are responsible for drawing legends, and the scale doesn't know how the lines should be labelled.

```
> huron <- data.frame(year = 1875:1972, level = LakeHuron)
> ggplot(huron, aes(year)) +
+   geom_line(aes(y = level - 5), colour = "blue") +
+   geom_line(aes(y = level + 5), colour = "red")
```

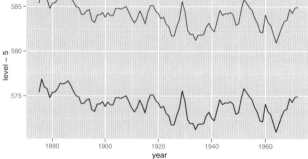

What you need to do is tell the colour scale about the two different lines by creating a mapping from the data to the colour aesthetic. There's no variable present in the data, so you'll have to create one:

```
> ggplot(huron, aes(year)) +
+   geom_line(aes(y = level - 5, colour = "below")) +
+   geom_line(aes(y = level + 5, colour = "above"))
```

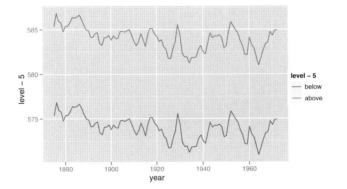

This gets us basically what we want, but the legend isn't labelled correctly, and has the wrong colours. That can be fixed with `scale_colour_manual`:

```
> ggplot(huron, aes(year)) +
+   geom_line(aes(y = level - 5, colour = "below")) +
+   geom_line(aes(y = level + 5, colour = "above")) +
+   scale_colour_manual("Direction",
+     c("below" = "blue", "above" = "red"))
```

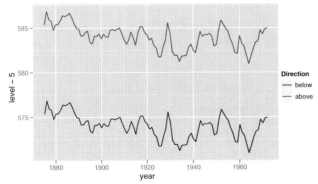

See Section 9.2.1 for an alternative approach to the problem.

6.4.5 The identity scale

The identity scale is used when your data is already in a form that the plotting functions in R understand, i.e., when the data and aesthetic spaces are the same. This means there is no way to derive a meaningful legend from the data alone, and by default a legend is not drawn. If you want one, you can still use the `breaks` and `labels` arguments to set it up yourself.

Figure 6.11 shows one sort of data where `scale_identity` is useful. Here the data themselves are colours, and there's no way we could make a meaningful legend. The identity scale can also be useful in the case where you have manually scaled the data to aesthetic values. In that situation, you will have to figure out what breaks and labels make sense for your data.

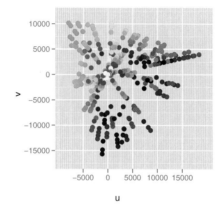

Fig. 6.11: A plot of R colours in Luv space. A legend is unnecessary, because the colour of the points represents itself: the data and aesthetic spaces are the same.

6.5 Legends and axes

Collectively, axes and legends are called **guides**, and they are the inverse of the scale: they allow you to read observations from the plot and map them back to their original values. Figure 6.12 labels the guides and their components. There are natural equivalents between the legend and the axis: the legend title and axis label are equivalent and determined by the scale name; the legend keys and tick labels are both determined by the scale breaks.

In ggplot2, legends and axes are produced automatically based on the scales and geoms that you used in the plot. This is different than how legends work in most other plotting systems, where you are responsible for adding them. In ggplot2, there is little you can do to directly control the legend. This seems like a big restriction at first, but as you get more comfortable with this approach, you will discover that it saves you a lot of time, and there is little you cannot do with it.

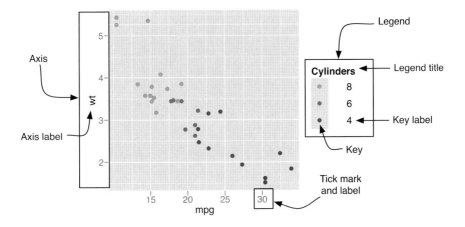

Fig. 6.12: The components of the axes and legend.

To draw the legend, the plot must collect information about how each aesthetic is used: for what data and what geoms. The scale breaks are used to determine the values of the legend keys and a list of the geoms that use the aesthetic is used to determine how to draw the keys. For example, if you use the point geom, then you will get points in the legend; if you use the lines geom, you will get lines. If both point and line geoms are used, then both points and lines will be drawn in the legend. This is illustrated in Figure 6.13.

Fig. 6.13: Legends produced by geom: point, line, point and line, and bar.

ggplot2 tries to use the smallest possible number of legends that accurately conveys the aesthetics used in the plot. It does this by combining legends if a variable is used with more than one aesthetic. Figure 6.14 shows an example of this for the points geom: if both colour and shape are mapped to the same variable, then only a single legend is necessary. In order for legends to be merged, they must have the same name (the same legend title). For this reason, if you change the name of one of the merged legends, you'll need to change it for all of them.

Fig. 6.14: Colour legend, shape legend, colour + shape legend.

The contents of the legend and axes are controlled by the scale, and the details of the rendering are controlled by the theming system. The following list includes the most commonly tweaked settings.

- The scale `name` controls the axis label and the legend title. This can be a string, or a mathematical expression, as described in `?plotmath`.
- The `breaks` and `labels` arguments to the scale function, introduced earlier in this chapter, are particularly important because they control what tick marks appear on the axis and what keys appear on the legend. If the breaks chosen by default are not appropriate (or you want to use more informative labels), setting these arguments will adjust the appearance of the legend keys and axis ticks.
- The theme settings `axis.*` and `legend.*` control the visual appearance of axes and legends. To learn how to manipulate these settings, see Section 8.1.
- The internal grid lines are controlled by the breaks and minor breaks arguments. By default minor grid lines are spaced evenly in the original data space: this gives the common behaviour of log-log plots where major grid lines are multiplicative and minor grid lines are additive. You can override the minor grid lines with the `minor_breaks` argument. Grid line appearance is controlled by the `panel.grid.major` and `panel.grid.minor` theme settings.
- The position and justification of legends are controlled by the theme setting `legend.position`, and the value can be right, left, top, bottom, none (no legend), or a numeric position. The numeric position gives (in values between 0 and 1) the position of the corner given by `legend.justification`, a numeric vector of length two. Top right = `c(1, 1)`, bottom left = `c(0, 0)`.

6.6 More resources

As you experiment with different aesthetic choices and new scales, it's important to keep in mind how the plot will be perceived. Some particularly good references to consult are:

- Cleveland (1993a, 1985); Cleveland and McGill (1987) for research on how plots are perceived and the best ways to encode data.
- Tufte (1990, 1997, 2001, 2006) for how to make beautiful, data-rich, graphics.
- Brewer (1994a,b) for how to choose colours that work well in a wide variety of situations, particularly for area plots.
- Carr (1994, 2002); Carr and Sun (1999) for the use of colour in general.

Chapter 7

Positioning

7.1 Introduction

This chapter discusses position, particularly how facets are laid out on a page, and how coordinate systems within a panel work. There are four components that control position. You have already learned about two of them that work within a facet:

- **Position adjustments** adjust the position of overlapping objects within a layer, and were described in Section 4.8. These are most useful for bar and other interval geoms, but can be useful in other situations.
- **Position scales**, previously described in Section 6.4.2, control how the values in the data are mapped to positions on the plot. Common transformations are linear and log, but any other invertible function can also be used.

This chapter will describe the other two components and show you how all four components can be used together:

- **Faceting**, described in Section 7.2, is a mechanism for automatically laying out multiple plots on a page. It splits the data into subsets, and then plots each subset into a different panel on the page. Such plots are often called small multiples.
- **Coordinate systems**, described in Section 7.3, control how the two independent position scales are combined to create a 2d coordinate system. The most common coordinate system is Cartesian, but other coordinate systems can be useful in special circumstances.

7.2 Faceting

You first encountered faceting in the introduction to `qplot()`, Section 2.6, and you may already have been using it in your plots. Faceting generates small

H. Wickham, *ggplot2*, Use R, DOI 10.1007/978-0-387-98141-3_7,
© Springer Science+Business Media, LLC 2009

multiples each showing a different subset of the data. Small multiples are a powerful tool for exploratory data analysis: you can rapidly compare patterns in different parts of the data and see whether they are the same or different. This section will discuss how you can fine-tune facets, particularly the way in which they interact with position scales.

There are two types of faceting provided by ggplot2: facet_grid and facet_wrap. Facet grid produces a 2d grid of panels defined by variables which form the rows and columns, while facet wrap produces a 1d ribbon of panels that is wrapped into 2d. The grid layout is similar to the layout of coplot in base graphics, and the wrapped layout is similar to the layout of panels in lattice. These differences are illustrated in Figure 7.1.

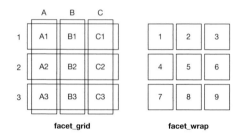

Fig. 7.1: A sketch illustrating the difference between the two faceting systems. facet_grid() (left) is fundamentally 2d, being made up of two independent components. facet_wrap() (right) is 1d, but wrapped into 2d to save space.

There are two basic arguments to the faceting systems: the variables to facet by, and whether position scales should be global or local to the facet. The way these options are specified is a little different for the two systems, so they are described separately below.

You can access either faceting system from qplot(). A 2d faceting specification (e.g., x ~ y) will use facet_grid, while a 1d specification (e.g., ~ x) will use facet_wrap.

Faceted plots have the capability to fill up a lot of space, so for this chapter we will use a subset of the mpg dataset that has a manageable number of levels: three cylinders (4, 6, 8) and two types of drive train (4 and f). This removes 29 vehicles from the original dataset.

```
> mpg2 <- subset(mpg, cyl != 5 & drv %in% c("4", "f"))
```

7.2.1 Facet grid

The grid faceter lays out plots in a 2d grid. When specifying a faceting formula, you specify which variables should appear in the columns and which should appear in the rows, as follows:

- . ˜ . The default. Neither rows nor columns are faceted, so you get a single panel.

```
> qplot(cty, hwy, data = mpg2) + facet_grid(. ~ .)
```

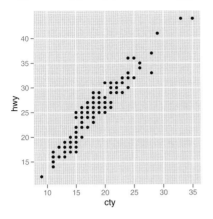

- . ˜ a A single row with multiple columns. This is normally the most useful direction because computer screens are usually wider than they are long. This direction of faceting facilitates comparisons of y position, because the vertical scales are aligned.

```
> qplot(cty, hwy, data = mpg2) + facet_grid(. ~ cyl)
```

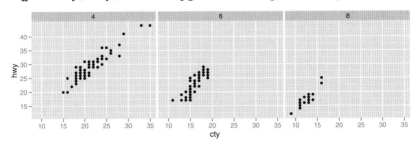

- b ˜ . A single column with multiple rows. This direction facilitates comparison of x position, because the horizontal scales are aligned, and so is particularly useful for comparing distributions. Figure 2.16 on page 24 is a good example of this use.

```
> qplot(cty, data = mpg2, geom="histogram", binwidth = 2) +
+    facet_grid(cyl ~ .)
```

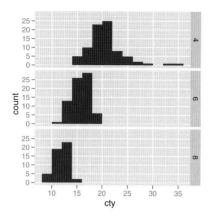

- **a ~ b**: Multiple rows and columns. You'll usually want to put the variable with the greatest number of levels in the columns, to take advantage of the aspect ratio of your screen.

```
> qplot(cty, hwy, data = mpg2) + facet_grid(drv ~ cyl)
```

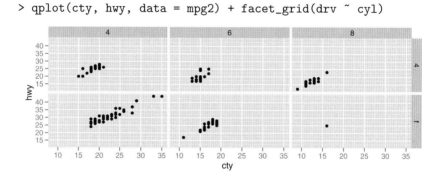

- **. ~ a + b** or **a + b ~ .** . Multiple variables in the rows or columns (or both). This is unlikely to be useful unless the number of factor levels is small, you have a very wide screens or you want to produce a long, skinny poster.

Variables appearing together on the rows or columns are nested in the sense that only combinations that appear in the data will appear in the plot. Variables that are specified on rows and columns will be crossed: all combinations will be shown, including those that didn't appear in the original dataset: this may result in empty panels.

Margins

Faceting a plot is like creating a contingency table. In contingency tables it is often useful to display marginal totals (totals over a row or column) as well as the individual cells. It is also useful to be able to do this with graphics, and you can do so with the `margins` argument. This allows you to compare the conditional patterns with the marginal patterns.

You can either specify that all margins should be displayed, using `margins = TRUE`, or by listing the names of the variables that you want margins for, `margins = c("sex", "age")`. You can also use `"grand_row"` or `"grand_col"` to produce grand row and grand column margins, respectively.

Figure 7.2 shows what margins look like. The first plot shows what the data looks like without margins, and the second shows all margins. The margin column shows all drive trains, the margin row shows all cylinders and the bottom right plot (the grand total) shows the full dataset. For this data we can see that as the number of cylinders increases, engine displacement increases and fuel economy decreases, and compared to front-wheel-drive vehicles, as a group four-wheel-drive vehicles have about the same displacement, but are less fuel efficient. The figure was produced with the following code:

```
p <- qplot(displ, hwy, data = mpg2) +
  geom_smooth(method = "lm", se = F)
p + facet_grid(cyl ~ drv)
p + facet_grid(cyl ~ drv, margins = T)
```

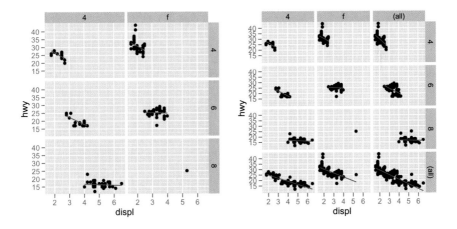

Fig. 7.2: Graphical margins work like margins of a contingency table to give unconditioned views of the data. A plot faceted by number of cylinders and drive train (left) is supplemented with margins (right).

Groups in the margins are controlled in the same way as groups in all other panels, defaulting to the interaction of all categorical variables present in the layer. (See Section 4.5.3 for a reminder.) The following example shows what happens when we add a coloured smooth for each drive train.

```
> qplot(displ, hwy, data = mpg2) +
+    geom_smooth(aes(colour = drv), method = "lm", se = F) +
+    facet_grid(cyl ~ drv, margins = T)
```

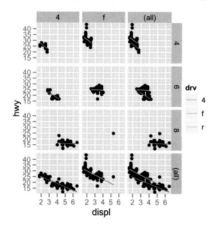

Plots with many facets and margins may be more appropriate for printing than on screen display, as the higher resolution of print (600 dpi vs. 72 dpi) allows you to compare many more subsets.

7.2.2 Facet wrap

An alternative to the grid is a wrapped ribbon of plots. Instead of having a 2d grid generated by the combination of two (or more) variables, facet_wrap makes a long ribbon of panels (generated by any number of variables) and wraps it into 2d. This is useful if you have a single variable with many levels and want to arrange the plots in a more space efficient manner. This is what trellising in lattice does.

Figure 7.3 shows the distribution of average movie ratings by decade. The main difference over time seems to be the increasing spread of ratings. This is probably an artefact of the number of votes: newer movies get more votes and so the average ratings are likely to be less extreme. The disadvantage of this style of faceting is that it is harder to compare some subsets that should be close together, as in this example where the plots for the 50's and 60's are particularly far apart because of the way the ribbon has been wrapped around. The figure was produced with the following code:

```
movies$decade <- round_any(movies$year, 10, floor)
qplot(rating, ..density.., data=subset(movies, decade > 1890),
   geom="histogram", binwidth = 0.5) +
   facet_wrap(~ decade, ncol = 6)
```

The specification of faceting variables is of the form ~ a + b + c. By default, facet_wrap will try and lay out the panels as close to a square as

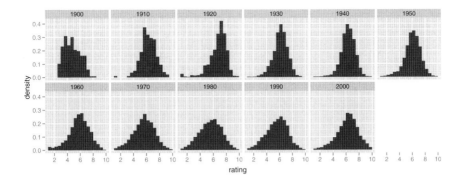

Fig. 7.3: Movie rating distribution by decade.

possible, with a slight bias towards wider rather than taller rectangles. You can override the default by setting `ncol`, `nrow` or both. See the documentation for more examples.

7.2.3 Controlling scales

For both types of faceting you can control whether the position scales are the same in all panels (fixed) or allowed to vary between panels (free). This is controlled by the `scales` parameter:

- `scales = "fixed"`: x and y scales are fixed across all panels.
- `scales = "free"`: x and y scales vary across panels.
- `scales = "free_x"`: the x scale is free, and the y scale is fixed.
- `scales = "free_y"`: the y scale is free, and the x scale is fixed.

Figure 7.4 illustrates the difference between the two extremes of fixed and free.

```
p <- qplot(cty, hwy, data = mpg)
p + facet_wrap(~ cyl)
p + facet_wrap(~ cyl, scales = "free")
```

Fixed scales allow us to compare subsets on an equal basis, seeing where each fits into the overall pattern. Free scales zoom in on the region that each subset occupies, allowing you to see more details. Free scales are particularly useful when we want to display multiple times series that were measured on different scales. To do this, we first need to change from "wide" to "long" data, stacking the separate variables into a single column. An example of this is shown in Figure 7.5, and the topic is discussed in more detail in Section 9.2.

```
em <- melt(economics, id = "date")
qplot(date, value, data = em, geom = "line", group = variable) +
    facet_grid(variable ~ ., scale = "free_y")
```

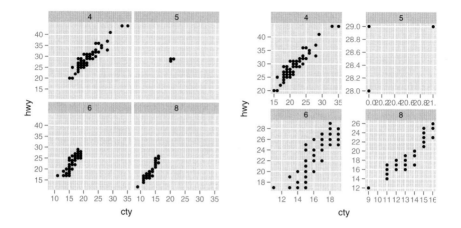

Fig. 7.4: Fixed scales (left) have the same scale for each facet, while free scales (right) have a different scale for each facet.

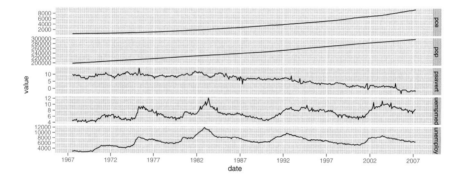

Fig. 7.5: Free scales are particularly useful when displaying multiple time series measured on different scales.

There is an additional constraint on the scales of `facet_grid`: all panels in a column must have the same x scale, and all panels in a row must have the same y scale. This is because each column shares an x axis, and each row shares a y axis.

For `facet_grid` there is an additional parameter called `space`, which takes values `"free"` or `"fixed"`. When the space can vary freely, each column (or row) will have width (or height) proportional to the range of the scale for that column (or row). This makes the scaling equal across the whole plot: 1 cm on each panel maps to the same range of data. (This is somewhat analogous to the "sliced" axis limits of lattice.) For example, if panel a had range 2 and panel b had range 4, one-third of the space would be given to a, and two-thirds to b. This is most useful for categorical scales, where we can assign space proportionally based on the number of levels in each facet, as illustrated by Figure 7.6. The code to create this plot is shown below: note the use of `reorder()` to arrange the models and manufacturers in order of city fuel usage.

```
mpg3 <- within(mpg2, {
  model <- reorder(model, cty)
  manufacturer <- reorder(manufacturer, -cty)
})
models <- qplot(cty, model, data = mpg3)

models
models + facet_grid(manufacturer ~ ., scales = "free",
  space = "free") +  opts(strip.text.y = theme_text())
```

7.2.4 Missing faceting variables

If you using faceting on a plot with multiple datasets, what happens when one of those datasets is missing the faceting variables? This situation commonly arises when you are adding contextual information that should be the same in all panels. For example, imagine you have spatial display of disease faceted by gender. What happens when you add a map layer that does not contain the gender variable? Here `ggplot2` will do what you expect: it will display the map in every facet: missing faceting variables are treated like they have all values.

7.2.5 Grouping vs. faceting

Faceting is an alternative to using aesthetics (like colour, shape or size) to differentiate groups. Both techniques have strengths and weaknesses, based around the relative positions of the subsets.

With faceting, each group is quite far apart in its own panel, and there is no overlap between the groups. This is good if the groups overlap a lot,

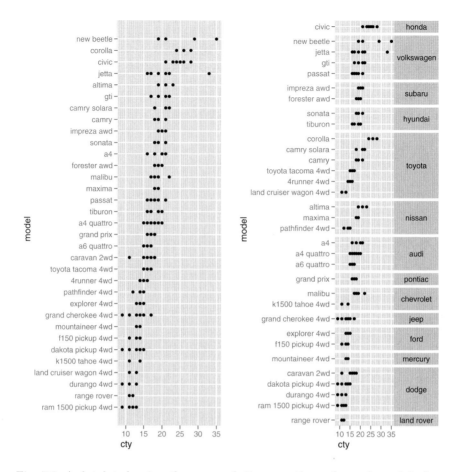

Fig. 7.6: A dotplot showing the range of city gas mileage for each model of car. (Left) Models ordered by average mpg, and (right) faceted by manufacturer with `scales="free_y"` and `space = "free"`. The `strip.text.y` theme setting has been used to rotate the facet labels.

but it does make small differences harder to see. When using aesthetics to differentiate groups, the groups are close together and may overlap, but small differences are easier to see. Figure 7.7 illustrates these trade-offs. With the scatterplots, it is possible to not realise the groups are overlapping when just colour is used to separate them, but with the regression lines they are too far apart to see that D, E and G are grouped together and J is farther away. The code to produce these figures is shown below.

```
xmaj <- c(0.3, 0.5, 1,3, 5)
xmin <- as.vector(outer(1:10, 10^c(-1, 0)))
ymaj <- c(500, 1000, 5000, 10000)
ymin <- as.vector(outer(1:10, 10^c(2,3,4)))
dplot <- ggplot(subset(diamonds, color %in% c("D","E","G","J")),
  aes(carat, price, colour = color)) +
  scale_x_log10(breaks = xmaj, labels = xmaj, minor = xmin) +
  scale_y_log10(breaks = ymaj, labels = ymaj, minor = ymin) +
  scale_colour_hue(limits = levels(diamonds$color)) +
  opts(legend.position = "none")

dplot + geom_point()
dplot + geom_point() + facet_grid(. ~ color)

dplot + geom_smooth(method = lm, se = F, fullrange = T)
dplot + geom_smooth(method = lm, se = F, fullrange = T) +
  facet_grid(. ~ color)
```

Faceting will also work with much larger number of groups, and because you can split in two dimensions, you can compare two variables simultaneously more easily than using two different aesthetics. The other advantage of faceting is that the scales can vary across panels, which is useful if the subsets occupy very different ranges.

7.2.6 Dodging vs. faceting

Faceting can achieve similar effects to dodging. Figure 7.8 shows how dodging and faceting can create plots that look remarkably similar. The main difference is the labelling: the faceted plot has colour labelled above and cut below; and the dodged plot has colour below and cut is not explicitly labelled. In this example, the labels in the faceted plot need some adjustment to display in a readable way, see the code below for details.

```
qplot(color, data=diamonds, geom = "bar", fill = cut,
  position="dodge")
qplot(cut, data = diamonds, geom = "bar", fill = cut) +
  facet_grid(. ~ color) +
  opts(axis.text.x = theme_text(angle = 90, hjust = 1, size = 8,
  colour = "grey50"))
```

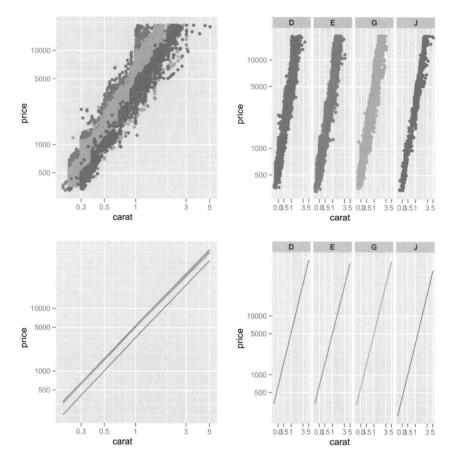

Fig. 7.7: The differences between faceting vs. grouping, illustrated with a log-log plot of carat vs. price with four selected colours.

Apart from labelling, the main difference between dodging and faceting occurs when the two variables are nearly completely crossed, but there are some combinations that do not occur. In this case, dodging becomes less useful because it only splits up the bars locally, and there are no labels. Faceting is more useful as we can control whether the splitting is local (`scales = "free_x"`, `space = "free"`) or global (`scales = "fixed"`). Figure 7.9 compares faceting and dodging for two nested variables from the `mpg` dataset, model and manufacturer, with the code shown below.

```
mpg4 <- subset(mpg, manufacturer %in%
  c("audi", "volkswagen", "jeep"))
mpg4$manufacturer <- as.character(mpg4$manufacturer)
mpg4$model <- as.character(mpg4$model)
```

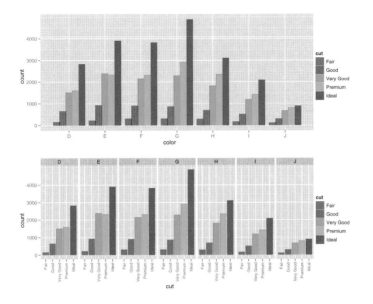

Fig. 7.8: Dodging (top) vs. faceting (bottom) for a completely crossed pair of variables.

```
base <- ggplot(mpg4, aes(fill = model)) +
  geom_bar(position = "dodge") +
  opts(legend.position = "none")

base + aes(x = model) +
  facet_grid(. ~ manufacturer)

last_plot() +
  facet_grid(. ~ manufacturer, scales = "free_x", space = "free")
base + aes(x = manufacturer)
```

In summary, the choice between faceting and dodging depends on the relationship between the two variables:

- Completely crossed: faceting and dodging are basically equivalent.
- Almost crossed: faceting with shared scales ensures that all combinations are visible, even if empty. This is particularly useful if missing combinations are non-structural missings.
- Nested: faceting with free scales and space allocates just enough space for each higher level group, and labels each item individually.

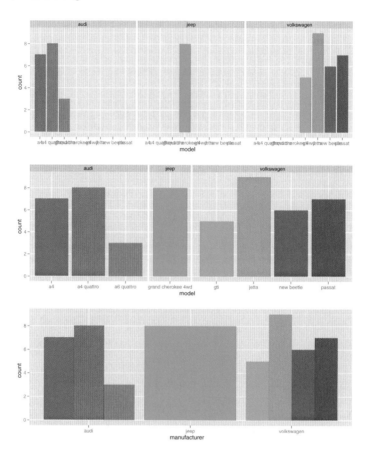

Fig. 7.9: For nested data, there is a clear advantage to faceting (top and middle) compared to dodging (bottom), because it is possible to carefully control and label the facets. For this example, the top plot is not useful, but it will be useful in situations where the data is almost crossed, i.e. where a single combination is missing.

7.2.7 Continuous variables

You can facet by continuous variables, but you will need to convert them into discrete categories first. There are three ways to do this:

- Divide the data into n bins each of the same length: `cut_interval(x, n = 10)` to specify the number of bins, or `cut_interval(x, length = 1)` to specify the length of each interval. Specifying the number of bins is easy, but may produce ranges that are not "nice" numbers.
- Divide the data into n bins each containing (approximately) the same number of points: `cut_number(x, n = 10)`. This makes it easier to compare facets (they will all have the same number of points), but you need to note that the range of each bin is different.

The following code demonstrates each of the three possibilities, with the results shown in Figure 7.10.

```
mpg2$disp_ww <- cut_interval(mpg2$displ, length = 1)
mpg2$disp_wn <- cut_interval(mpg2$displ, n = 6)
mpg2$disp_nn <- cut_number(mpg2$displ, n = 6)

plot <- qplot(cty, hwy, data = mpg2) + labs(x = NULL, y = NULL)
plot + facet_wrap(~ disp_ww, nrow = 1)
plot + facet_wrap(~ disp_wn, nrow = 1)
plot + facet_wrap(~ disp_nn, nrow = 1)
```

Note that the faceting formula only works with variables in the dataset (not functions of the variables), so you will also need to create a new variable containing the discretised data.

7.3 Coordinate systems

Coordinate systems tie together the two position scales to produce a 2d location. Currently, `ggplot2` comes with six different coordinate systems, listed in Table 7.1. All these coordinate systems are two dimensional, although one day I hope to add 3d graphics too. As with the other components in `ggplot2`, you generate the R name by joining `coord_` and the name of the coordinate system. Most plots use the default Cartesian coordinate system, `coord_cartesian()`, where the 2d position of an element is given by the combination of the x and y positions.

Coordinate systems have two main jobs:

- Combine the two position aesthetics to produce a 2d position on the plot. The position aesthetics are called x and y, but they might be better called position 1 and 2 because their meaning depends on the coordinate system used. For example, with the polar coordinate system they become angle

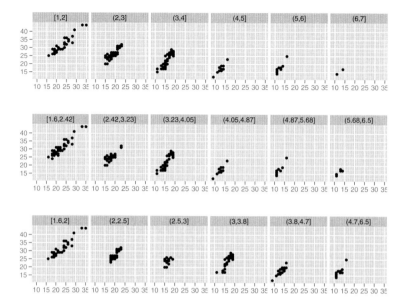

Fig. 7.10: Three ways of breaking a continuous variable into discrete bins. From top to bottom: bins of length one, six bins of equal length, six bins containing equal numbers of points.

and radius (or radius and angle), and with maps they become latitude and longitude.

- In coordination with the faceter, coordinate systems draw axes and panel backgrounds. While the scales control the values that appear on the axes, and how they map from data to position, it is the coordinate system which actually draws them. This is because their appearance depends on the coordinate system: an angle axis looks quite different than an x axis.

7.3.1 Transformation

Unlike transforming the data or transforming the scales, transformations carried out by the coordinate system change the appearance of the geoms: in polar coordinates a rectangle becomes a slice of a doughnut; in a map projection, the shortest path between two points will no longer be a straight line. Figure 7.11 illustrates what happens to a line and a rectangle in a few different coordinate systems.

This transformation takes part in two steps. Firstly, the parameterisation of each geom is changed to be purely location-based, rather than location and dimension based. For example, a bar can be represented as an x position (a location), a height and a width (two dimensions). But how do we interpret

Name	Description
cartesian	Cartesian coordinates
equal	Equal scale Cartesian coordinates
flip	Flipped Cartesian coordinates
trans	Transformed Cartesian coordinate system
map	Map projections
polar	Polar coordinates

Table 7.1: Coordinate systems available in ggplot. `coord_equal`, `coord_flip` and `coord_trans` are all basically Cartesian in nature (i.e., the dimensions combine orthogonally), while `coord_map` and `coord_polar` are more complex.

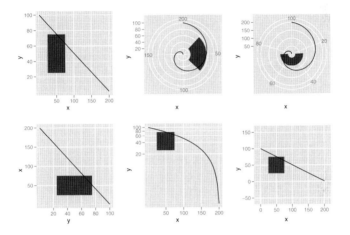

Fig. 7.11: A set of examples illustrating what a line and rectangle look like when displayed in a variety of coordinate systems. From top left to bottom right: Cartesian, polar with x position mapped to angle, polar with y position mapped to angle, flipped, transformed with log in y direction, and equal scales.

height and width in a non-Cartesian coordinate system, where a rectangle may not have constant height and width? We solve the problem by using a purely location-based representation, the location of the four corners of the rectangle, and then transforming these locations: we have converted a rectangle to a polygon. By doing this, we effectively convert all geoms to a combination of points, lines and polygons.

With all geoms in this consistent, location-based, representation, the next step is to transform each location into the new coordinate system. It is easy to transform points, because a point is still a point no matter what coordinate system you are in, but lines and polygons are harder, because a straight line may no longer be straight in the new coordinate system. To make the problem tractable we assume that all coordinate transformations are smooth, in the sense that all very short lines will still be very short straight lines in the new coordinate system. With this assumption in hand, we can transform lines and polygons by breaking them up into many small line segments and transforming each segment. This process is called munching. Figure 7.12 illustrates this procedure. We start with a line parameterised by its two endpoints, then break it into multiple line segments, each with two endpoints. Those points are then translated into the new coordinate system, and connected. In the example, the number of line segments is too small, so you can see more easily how it works. For practical use, we use many more segments so that the result looks smooth.

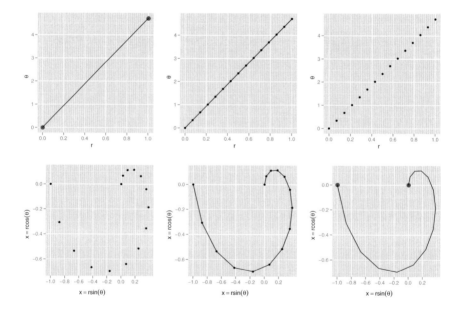

Fig. 7.12: How coordinate transformations work: converting a line in Cartesian coordinates to a line in polar coordinates. The original x position is converted to radius, and the y position to angle.

7.3.2 Statistics

To be technically correct, the actual statistical method used by a stat should depend on the coordinate system. For example, a smoother in polar coordinates should use circular regression, and in 3d should return a 2d surface rather than a 1d curve. However, many statistical operations have not been derived for non-Cartesian coordinates and `ggplot2` falls back to Cartesian coordinates for calculation, which, while not strictly correct, will normally be a fairly close approximation.

7.3.3 Cartesian coordinate systems

The four Cartesian-based coordinate systems, `coord_cartesian`, `coord_equal`, `coord_flip` and `coord_trans`, share a number of common features. They are still essentially Cartesian because the x and y positions map orthogonally to x and y positions on the plot.

Setting limits.

`coord_cartesian` has arguments `xlim` and `ylim`. If you think back to the scales chapter, you might wonder why we need these. Doesn't the limits argument of the scales already allow use to control what appears on the plot? The key difference is how the limits work: when setting scale limits, any data outside the limits is thrown away; but when setting coordinate system limits we still use all the data, but we only display a small region of the plot. Setting coordinate system limits is like looking at the plot under a magnifying glass. Figures 7.13 and 7.14 show an example of this.

```
(p <- qplot(disp, wt, data=mtcars) + geom_smooth())
p + scale_x_continuous(limits = c(325, 500))
p + coord_cartesian(xlim = c(325, 500))

(d <- ggplot(diamonds, aes(carat, price)) +
  stat_bin2d(bins = 25, colour="grey70") +
  opts(legend.position = "none"))
d + scale_x_continuous(limits = c(0, 2))
d + coord_cartesian(xlim = c(0, 2))
```

Flipping the axes.

Most statistics and geoms assume you are interested in y values conditional on x values (e.g., smooth, summary, boxplot, line): in most statistical models, the x values are assumed to be measured without error. If you are interested in x conditional on y (or you just want to rotate the plot 90 degrees), you can use `coord_flip` to exchange the x and y axes. Compare this with just exchanging the variables mapped to x and y, as shown in Figure 7.15.

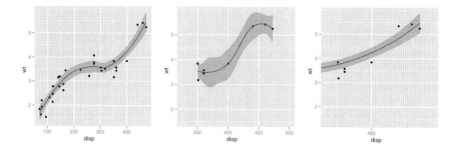

Fig. 7.13: Setting limits on the coordinate system, vs setting limits on the scales. (Left) Entire dataset; (middle) x scale limits set to (325, 500); (right) coordinate system x limits set to (325, 500). Scaling the coordinate limits performs a visual zoom, while setting the scale limits subsets the data and refits the smooth.

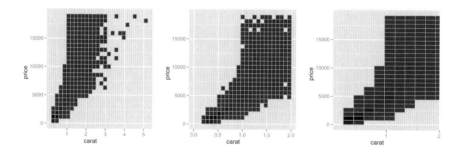

Fig. 7.14: Setting limits on the coordinate system, vs. setting limits on the scales. (Left) Entire dataset; (middle) x scale limits set to (0, 2); (right) coordinate x limits set to (0, 2). Compare the size of the bins: when you set the scale limits, there are the same number of bins but they each cover a smaller region of the data; when you set the coordinate limits, there are fewer bins and they cover the same amount of data as the original.

```
qplot(displ, cty, data = mpg) + geom_smooth()
qplot(cty, displ, data = mpg) + geom_smooth()
qplot(cty, displ, data = mpg) + geom_smooth() + coord_flip()
```

Transformations.

Like limits, we can also transform the data in two places: at the scale level or at the coordinate system level. `coord_trans` has arguments x and y which should be strings naming the transformer (Table 6.2) to use for that axis. Transforming at the scale level occurs before statistics are computed and does not change the shape of the geom. Transforming at the coordinate system level occurs after the statistics have been computed, and does affect the shape of the geom. Using both together allows us to model the data on a transformed scale

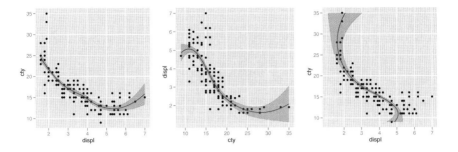

Fig. 7.15: (Left) A scatterplot and smoother with engine displacement on x axis and city mpg on y axis. (Middle) Exchanging cty and displ rotates the plot 90 degrees, but the smooth is fit to the rotated data. (Right) using coord_flip fits the smooth to the original data, and then rotates the output, this is a smooth curve of x conditional on y.

and then backtransform it for interpretation: a common pattern in analysis. An example of this is shown in Figure 7.16.

```
qplot(carat, price, data = diamonds, log = "xy") +
  geom_smooth(method = "lm")
last_plot() + coord_trans(x = "pow10", y = "pow10")
```

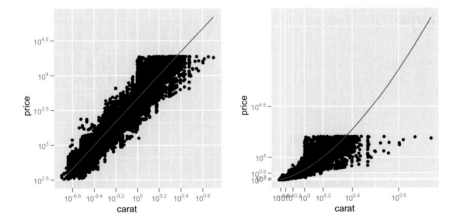

Fig. 7.16: (Left) A scatterplot of carat vs. price on log base 10 transformed scales. A linear regression summarises the trend: $\log(y) = a + b * \log(x)$. (Right) The previous plot backtransformed (with coord_trans(x = "pow10", y = "pow10")) onto the original scales. The linear trend line now becomes geometric, $y = k * c^x$, and highlights the lack of expensive diamonds for larger carats.

Equal scales.

`coord_equal` ensures that the x and y axes have equal scales: i.e., 1 cm along the x axis represents the same range of data as 1 cm along the y axis. By default it will assume that you want a one-to-one ratio, but you can change this with the `ratio` parameter. The aspect ratio will also be set to ensure that the mapping is maintained regardless of the shape of the output device. See the documentation of `coord_equal()` for more details.

7.3.4 Non-Cartesian coordinate systems

There are two non-Cartesian coordinate systems: polar coordinates and map projections. These coordinate systems are still somewhat experimental, and there are fewer standards for the layout of axes, so you may need to tweak them to meet your needs using the tools in Chapter C.

Polar coordinates.

Using polar coordinates gives rise to pie charts and wind roses (from bar geoms), and radar charts (from line geoms). Polar coordinates are often used for circular data, particularly time or direction, but the perceptual properties are not good because the angle is harder to perceive for small radii than it is for large radii. The `theta` argument determines which position variable is mapped to angle (by default, x) and which to radius. Figure 7.17 shows how by changing the coordinate system we can turn a bar chart into a pie chart or a bullseye chart. The documentation includes other examples of polar charts.

```
# Stacked barchart
(pie <- ggplot(mtcars, aes(x = factor(1), fill = factor(cyl))) +
  geom_bar(width = 1))
# Pie chart
pie + coord_polar(theta = "y")

# The bullseye chart
pie + coord_polar()
```

Map projections.

These are still rather experimental, and rely on the `mapproj` package (McIlroy, 2005). `coord_map()` takes the same arguments as `mapproj()` for controlling the projection. See the documentation of `coord_map()` for more examples, and consult a cartographer for the most appropriate projection for your data.

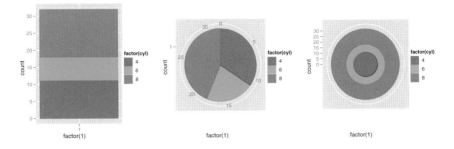

Fig. 7.17: (Left) A stacked bar chart. (Middle) The stacked bar chart in polar coordinates, with x position mapped to radius and y position mapped to angle, `coord_polar(theta = "y")`. This is more commonly known as a pie chart. (Right) The stacked bar chart in polar coordinates with the opposite mapping, `coord_polar(theta = "x")`. This is sometimes called a bullseye chart.

Chapter 8

Polishing your plots for publication

In this chapter you will learn how to prepare polished plots for publication. Most of this chapter focusses on the theming capability of `ggplot2` which allows you to control many non-data aspects of plot appearance, but you will also learn how to adjust geom, stat and scale defaults, and the best way to save plots for inclusion into other software packages. Together with the next chapter, manipulating plot rendering with `grid`, you will learn how to control every visual aspect of the plot to get exactly the appearance that you want.

The visual appearance of the plot is determined by both data and non-data related components. Section 8.1 introduces the theme system which controls all aspects of non-data display. By now you should be familiar with the many ways that you can alter the data-related components of the plot—layers and scales— to visualise your data and change the appearance of the plot. In Section 8.2 you will learn how you can change the defaults for these, so that you do not need to repeat the same parameters again and again.

Section 8.3 discusses the chapter with a discussion about how to get your graphics out of R and into LaTeX, Word or other presentation or word-processing software. Section 8.4 concludes with a discussion of how to lay out multiple plots on a single page.

8.1 Themes

The appearance of non-data elements of the plot is controlled by the theme system. The theme system does not affect how the data is rendered by geoms, or how it is transformed by scales. Themes don't change the perceptual properties of the plot, but they do help you make the plot aesthetically pleasing or match existing style guides. Themes give you control over things like the fonts in all parts of the plot: the title, axis labels, axis tick labels, strips, legend labels and legend key labels; and the colour of ticks, grid lines and backgrounds (panel, plot, strip and legend).

This separation of control into data and non-data parts is quite different than base and lattice graphics. In base and lattice graphics, most functions take

H. Wickham, *ggplot2*, Use R, DOI 10.1007/978-0-387-98141-3_8,
© Springer Science+Business Media, LLC 2009

a large number of arguments that specify both data and non-data appearance, which makes the functions complicated and hard to learn. ggplot2 takes a different approach: when creating the plot you determine how the data is displayed, then *after* it has been created you can edit every detail of the rendering, using the theming system. Some of the effects of changing the theme of a plot are shown in Figure 8.1. The two plots show the two themes included by default in ggplot2.

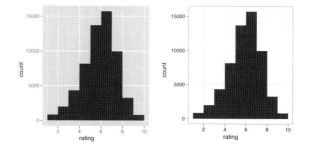

Fig. 8.1: The effect of changing themes. (Left) The default grey theme with grey background and white gridlines. (Right) the alternative black and white theme with white background and grey gridlines. Notice how the bars, data elements, are identical in both plots.

Like many other areas of ggplot2, themes can be controlled on multiple levels from the coarse to fine. You can:

- Use a built-in theme, as described in Section 8.1.1. This affects every element of the plot in a visually consistent manner. The default theme uses a grey panel background with white gridlines, while the alternative theme uses a white background with grey gridlines.
- Modify a single element of a built-in theme, as described in Section 8.1.2. Each theme is made up of multiple elements. The theme system comes with a number of built-in element rendering functions with a limited set of parameters. By adjusting these parameters you can control things like text size and colour, background and grid line colours and text orientation. By combining multiple elements you can create your own theme.

Generally each of these theme settings can be applied globally, to all plots, or locally to a single plot. How to do this is described in each section.

8.1.1 Built-in themes

There are two built-in themes. The default, theme_gray(), uses a very light grey background with white gridlines. This follows from the advice of Tufte

(1990, 1997, 2001, 2006) and Brewer (1994a); Carr (1994, 2002); Carr and Sun (1999). We can still see the gridlines to aid in the judgement of position (Cleveland, 1993b), but they have little visual impact and we can easily "tune" them out. The grey background gives the plot a similar colour (in a typographical sense) to the remainder of the text, ensuring that the graphics fit in with the flow of a text without jumping out with a bright white background. Finally, the grey background creates a continuous field of colour which ensures that the plot is perceived as a single visual entity.

The other built-in theme, `theme_bw()`, has a more traditional white background with dark grey gridlines. Figure 8.1 shows some of the difference between these themes.

Both themes have a single parameter, `base_size`, which controls the base font size. The base font size is the size that the axis titles use: the plot title is 20% bigger, and the tick and strip labels are 20% smaller. If you want to control these sizes separately, you'll need to modify the individual elements as described in the following section.

You can apply themes in two ways:

- Globally, affecting all plots when they are drawn: `theme_set(theme_grey())` or `theme_set(theme_bw())`. `theme_set()` returns the previous theme so that you can restore it later if you want.
- Locally, for an individual plot: `qplot(...) + theme_grey()`. A locally applied theme will override the global default.

The following example shows a few of these combinations:

```
> hgram <- qplot(rating, data = movies, binwidth = 1)
>
> # Themes affect the plot when they are drawn,
> # not when they are created
> hgram
```

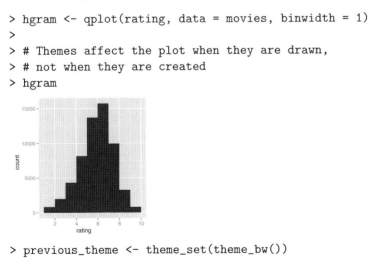

```
> previous_theme <- theme_set(theme_bw())
> hgram
```

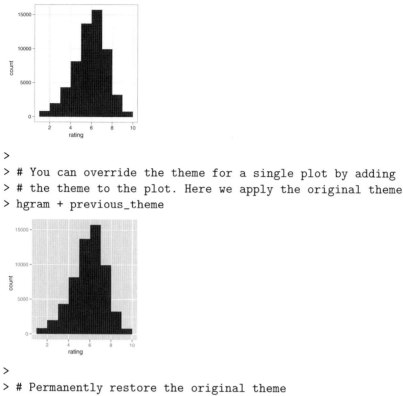

```
>
> # You can override the theme for a single plot by adding
> # the theme to the plot. Here we apply the original theme
> hgram + previous_theme
```

```
>
> # Permanently restore the original theme
> theme_set(previous_theme)
```

8.1.2 Theme elements and element functions

A theme is made up of multiple *elements* which control the appearance of a single item on the plot, as listed in Table 8.1. There are three elements that have individual x and y settings: `axis.text`, `axis.title` and `strip.text`. Having a different setting for the horizontal and vertical elements allows you to control how text should appear in different orientations. The appearance of each element is controlled by an *element function.*

There are four basic types of built-in element functions: text, lines and segments, rectangles and blank. Each element function has a set of parameters that control the appearance as described below:

- `theme_text()` draws labels and headings. You can control the font `family`, face, colour, size, hjust, vjust, angle and `lineheight`.
  The following code shows the effect of changing these parameters on the plot title. The results are shown in Figure 8.2. Changing the angle is probably more useful for tick labels. When changing the angle you will probably also need to change hjust to 0 or 1.

Theme element	Type	Description
`axis.line`	segment	line along axis
`axis.text.x`	text	x axis label
`axis.text.y`	text	y axis label
`axis.ticks`	segment	axis tick marks
`axis.title.x`	text	horizontal tick labels
`axis.title.y`	text	vertical tick labels
`legend.background`	rect	background of legend
`legend.key`	rect	background underneath legend keys
`legend.text`	text	legend labels
`legend.title`	text	legend name
`panel.background`	rect	background of panel
`panel.border`	rect	border around panel
`panel.grid.major`	line	major grid lines
`panel.grid.minor`	line	minor grid lines
`plot.background`	rect	background of the entire plot
`plot.title`	text	plot title
`strip.background`	rect	background of facet labels
`strip.text.x`	text	text for horizontal strips
`strip.text.y`	text	text for vertical strips

Table 8.1: Theme elements

```
hgramt <- hgram +
  opts(title = "This is a histogram")
hgramt
hgramt + opts(plot.title = theme_text(size = 20))
hgramt + opts(plot.title = theme_text(size = 20,
  colour = "red"))
hgramt + opts(plot.title = theme_text(size = 20,
  hjust = 0))
hgramt + opts(plot.title = theme_text(size = 20,
  face = "bold"))
hgramt + opts(plot.title = theme_text(size = 20,
  angle = 180))
```

- `theme_line()` and `theme_segment()` draw lines and segments with the same options but in a slightly different way. Make sure you match the appropriate type or you will get strange grid errors. For these element functions you can control the `colour`, `size` and `linetype`. These options are illustrated with the code and the results are shown in Figure 8.3.

```
hgram + opts(panel.grid.major = theme_line(colour = "red"))
```

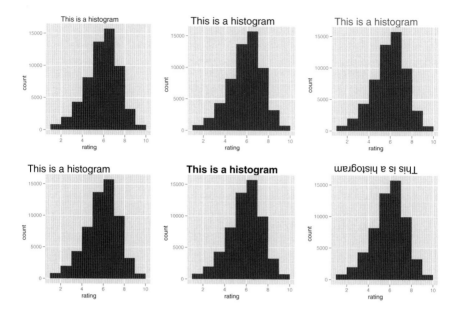

Fig. 8.2: Changing the appearance of the plot title.

```
hgram + opts(panel.grid.major = theme_line(size = 2))
hgram + opts(panel.grid.major = theme_line(linetype = "dotted"))
hgram + opts(axis.line = theme_segment())
hgram + opts(axis.line = theme_segment(colour = "red"))
hgram + opts(axis.line = theme_segment(size = 0.5,
  linetype = "dashed"))
```

- theme_rect() draws rectangles, mostly used for backgrounds, you can control the fill colour and border colour, size and linetype. Examples shown in Figure 8.4 are created with the code below:

```
hgram + opts(plot.background = theme_rect(fill = "grey80",
  colour = NA))
hgram + opts(plot.background = theme_rect(size = 2))
hgram + opts(plot.background = theme_rect(colour = "red"))
hgram + opts(panel.background = theme_rect())
hgram + opts(panel.background = theme_rect(colour = NA))
hgram + opts(panel.background =
  theme_rect(linetype = "dotted"))
```

- theme_blank() draws nothing. Use this element type if you don't want anything drawn, and no space allocated for that element. The following example uses theme_blank() to progressively suppress the appearance of elements we're not interested in. The results are shown in Figure 8.5. Notice

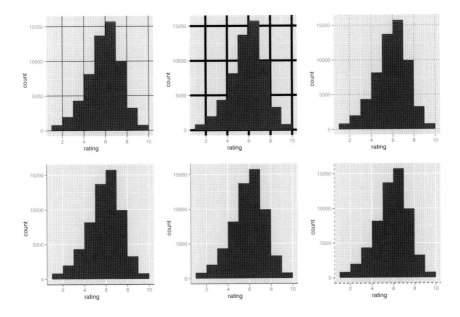

Fig. 8.3: Changing the appearance of lines and segments in the plot.

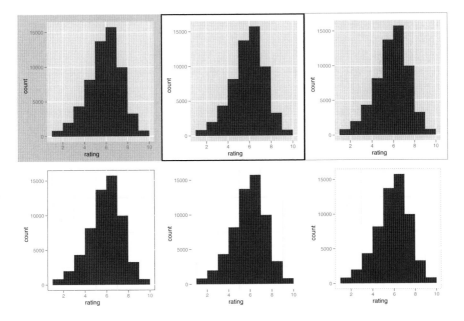

Fig. 8.4: Changing the appearance of the plot and panel background

how the plot automatically reclaims the space previously used by these
elements: if you don't want this to happen (perhaps because they need to
line up with other plots on the page), use colour = NA, fill = NA as
parameter to create invisible elements that still take up space.

```
hgramt
last_plot() + opts(panel.grid.minor = theme_blank())
last_plot() + opts(panel.grid.major = theme_blank())
last_plot() + opts(panel.background = theme_blank())
last_plot() +
  opts(axis.title.x = theme_blank(),
       axis.title.y = theme_blank())
last_plot() + opts(axis.line = theme_segment())
```

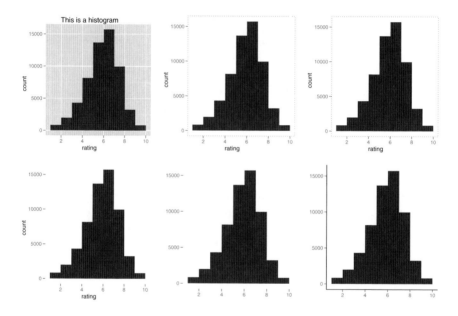

Fig. 8.5: Progressively removing non-data elements from a plot with
theme_blank().

You can see the settings for the current theme with theme_get(). The
output isn't included here because it takes up several pages. You can modify
the elements locally for a single plot with opts() (as seen above), or globally
for all future plots with theme_update(). Figure 8.6 shows the results of
pulling together multiple theme settings with the following code.

```
old_theme <- theme_update(
  plot.background = theme_rect(fill = "#3366FF"),
```

```
  panel.background = theme_rect(fill = "#003DF5"),
  axis.text.x = theme_text(colour = "#CCFF33"),
  axis.text.y = theme_text(colour = "#CCFF33", hjust = 1),
  axis.title.x = theme_text(colour = "#CCFF33", face = "bold"),
  axis.title.y = theme_text(colour = "#CCFF33", face = "bold",
    angle = 90)
)
qplot(cut, data = diamonds, geom="bar")
qplot(cty, hwy, data = mpg)
theme_set(old_theme)
```

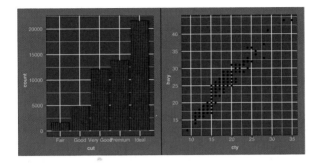

Fig. 8.6: A bar chart and scatterplot created after a new visually consistent (if ugly!) theme has been applied.

There is some duplication in this example because we have to specify the x and y elements separately. This is a necessary evil so that you can have total control over the appearance of the elements. If you are writing your own theme, you would probably want to write a function to minimise this repetition.

8.2 Customising scales and geoms

When producing a consistent theme, you may also want to tune some of the scale and geom defaults. Rather than having to manually specify the changes every time you add the scale or geom, you can use the following functions to alter the default settings for scales and geoms.

8.2.1 Scales

To change the default scale associated with an aesthetic, use `set_default_scale()`. (See Table 6.1 for the defaults.) This function takes three arguments: the name of the aesthetic, the type of variable (discrete or continuous) and the name of the scale to use as the default. Further arguments override the default parameters of the scale. The following example sets up colour and fill scales for black-and-white printing:

```
set_default_scale("colour", "discrete", "grey")
set_default_scale("fill", "discrete", "grey")
set_default_scale("colour", "continuous", "gradient",
  low = "white", high = "black")
set_default_scale("fill", "continuous", "gradient",
  low = "white", high = "black")
```

8.2.2 Geoms and stats

You can customise geoms and stats in a similar way with `update_geom_defaults()` and `update_stat_defaults()`. Unlike the other theme settings these will only affect plots *created* after the setting has been changed, not all plots drawn after the setting has been changed. The following example demonstrates changing the default point colour and changing the default histogram to a density ("true") histogram.

```
update_geom_defaults("point", aes(colour = "darkblue"))
qplot(mpg, wt, data=mtcars)
update_stat_defaults("bin", aes(y = ..density..))
qplot(rating, data = movies, geom = "histogram", binwidth = 1)
```

Table 8.2 lists all of the common aesthetic defaults. If you change the defaults for one geom, it's a good idea to change all the defaults for all the other geoms that you commonly use so that your plots look consistent. If you are unsure on what makes for a valid colour, line type, shape or size, Appendix B gives the details.

8.3 Saving your output

You have two basic choices of output: raster or vector. Vector graphics are procedural. This means that they are essentially "infinitely" zoomable; there is no loss of detail. Raster graphics are stored as an array of pixels and have a fixed optimal viewing size. Figure 8.7 illustrates the basic differences for a basic circle. A good description is available at `http://tinyurl.com/rstrvctr`.

Generally, vector output is more desirable, but for complex graphics containing thousands of graphical objects it can be slow to render. In this case, it may be better to switch to raster output. For printed use, a high-resolution (e.g., 600 dpi) graphic may be an acceptable compromise, but may be large.

To save your output, you can use the typical R way with disk-based graphics devices, which works for all packages, or a special function from `ggplot2` that saves the current plot: `ggsave()`. `ggsave()` is optimised for interactive use and has the following important arguments:

Aesthetic	Default value	Geoms
colour	#3366FF	contour, density2d, quantile, smooth
colour	NA	area, bar, histogram, polygon, rect, tile
colour	black	abline, crossbar, density, errorbar, hline, line, linerange, path, pointrange, rug, segment, step, text, vline
colour	darkblue	jitter, point
colour	grey60	boxplot, ribbon
fill	NA	crossbar, density, jitter, point, pointrange
fill	grey20	area, bar, histogram, polygon, rect, ribbon, tile
linetype	1	abline, area, bar, contour, crossbar, density, density2d, errorbar, histogram, hline, line, linerange, path, pointrange, polygon, quantile, rect, ribbon, rug, segment, smooth, step, tile, vline
shape	19	jitter, point, pointrange
size	0.5	abline, area, bar, boxplot, contour, crossbar, density, density2d, errorbar, histogram, hline, line, linerange, path, pointrange, polygon, quantile, rect, ribbon, rug, segment, smooth, step, vline
size	2	jitter, point
weight	1	bar, boxplot, contour, density, density2d, histogram, quantile, smooth

Table 8.2: Default aesthetic values for geoms. See Appendix B for how the values are interpreted by R.

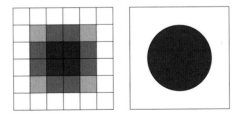

Fig. 8.7: The schematic difference between raster (left) and vector (right) graphics.

- The `path` specifies the path where the image should be saved. The file extension will be used to automatically select the correct graphics device.
- Three arguments control output size. If left blank, the size of the current on-screen graphics device will be used. `width` and `height` can be used to specify the absolute size, or `scale` to specify the size of the plot relative to the on-screen display. When creating the final versions of graphics it's a good idea to set `width` and `height` so you know exactly what size output you're going to get.
- For raster graphics, the `dpi` argument controls the resolution of the plot. It defaults to 300, which is appropriate for most printers, but you may want to use 600 for particularly high-resolution output, or 72 for on-screen (e.g., web) display.

The following code shows these two methods. If you want to save multiple plots to a single file, you will need to explicitly open a disk-based graphics device (like `png()` or `pdf()`), print the plots and then close it with `dev.off()`.

```
qplot(mpg, wt, data = mtcars)
ggsave(file = "output.pdf")

pdf(file = "output.pdf", width = 6, height = 6)
# If inside a script, you will need to explicitly print() plots
qplot(mpg, wt, data = mtcars)
qplot(wt, mpg, data = mtcars)
dev.off()
```

Table 8.3 lists recommended graphic formats for various tasks. R output generally works best as part of a linux development tool chain: using png or pdf output in LaTeX documents. With Microsoft Office it is easiest to use a high-resolution (`dpi = 600`) png file. You can use vector output, but neither Windows meta files nor postscript supports transparency, and while postscript prints fine, it is only shown on screen if you add a preview in another software package. Transparency is used to show confidence intervals with the points showing through. If you copy and paste a graph into Word, and see that the confidence interval bands have vanished, that is the cause. The same advice holds for OpenOffice.

If you are using LaTeX, I recommend including `\DeclareGraphicsExtensions{.png,.pdf}` in the preamble. Then you don't need to specify the file extension in `includegraphics` commands, but LaTeX will pick png files in preference to pdf. I choose this order because you can produce all your files in pdf, and then go back and re-render any big ones as png. Another useful command is `\graphicspath{{include/}}` which specifies a path in which to look for graphics, allowing you to keep graphics in a separate directory to the text.

Software	Recommended graphics device
Illustrator	svg
latex	ps
MS Office	png (600 dpi)
Open Office	png (600 dpi)
pdflatex	pdf, png (600 dpi)
web	png (72 dpi)

Table 8.3: Recommended graphic output for different purposes.

8.4 Multiple plots on the same page

If you want to arrange multiple plots on a single page, you'll need to learn a little bit of grid, the underlying graphics system used by `ggplot2`. The key concept you'll need to learn about is a viewport: a rectangular subregion of the display. The default viewport takes up the entire plotting region, and by customising the viewport you can arrange a set of plots in just about any way you can imagine.

To begin, let's create three plots that we can experiment with. When arranging multiple plots on a page, it will usually be easiest to create them, assign them to variables and then plot them. This makes it easier to experiment with plot placement independent of content. The plots created by the code below are shown in Figure 8.8.

```
(a <- qplot(date, unemploy, data = economics, geom = "line"))
(b <- qplot(uempmed, unemploy, data = economics) +
  geom_smooth(se = F))
(c <- qplot(uempmed, unemploy, data = economics, geom="path"))
```

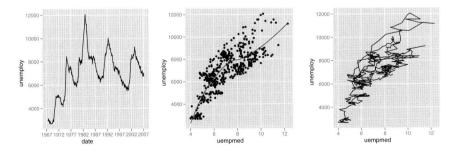

Fig. 8.8: Three simple graphics we'll use to experiment with sophisticated plot layouts.

8.4.1 Subplots

One common layout is to have a small subplot embedded drawn on top of the main plot. To achieve this effect, we first plot the main plot, and then draw the subplot in a smaller viewport. Viewports are created with (surprise!) the viewport() function, with parameters x, y, width and height to control the size and position of the viewport. By default, the measurements are given in "npc" units, which range from 0 to 1. The location (0, 0) is the bottom left, (1, 1) the top right and (0.5, 0.5) the centre of viewport. If these relative units don't work for your needs, you can also use absolute units, like unit(2, "cm") or unit(1, "inch").

```
# A viewport that takes up the entire plot device
vp1 <- viewport(width = 1, height = 1, x = 0.5, y = 0.5)
vp1 <- viewport()

# A viewport that takes up half the width and half the height,
# located in the middle of the plot.
vp2 <- viewport(width = 0.5, height = 0.5, x = 0.5, y = 0.5)
vp2 <- viewport(width = 0.5, height = 0.5)

# A viewport that is 2cm x 3cm located in the center
vp3 <- viewport(width = unit(2, "cm"), height = unit(3, "cm"))
```

By default, the x and y parameters control the location of the centre of the viewport. When positioning the plot in other locations, you may need to use the just parameter to control which corner of the plot you are positioning. The following code gives some examples.

```
# A viewport in the top right
vp4 <- viewport(x = 1, y = 1, just = c("top", "right"))
# Bottom left
vp5 <- viewport(x = 0, y = 0, just = c("bottom", "right"))
```

To draw the plot in our new viewport, we use the vp argument of the ggplot print() method. This method is normally called automatically whenever you evaluate something on the command line, but because we want to customise the viewport, we need to call it ourselves. The result of this is shown in Figure 8.9(a).

```
pdf("polishing-subplot-1.pdf", width = 4, height = 4)
subvp <- viewport(width = 0.4, height = 0.4, x = 0.75, y = 0.35)
b
print(c, vp = subvp)
dev.off()
```

This gives us what we want, but we need to make a few tweaks to the appearance: the text should be smaller, we want to remove the axis labels and shrink the plot margins. The result is shown in Figure 8.9(b).

```
csmall <- c +
  theme_gray(9) +
  labs(x = NULL, y = NULL) +
  opts(plot.margin = unit(rep(0, 4), "lines"))

pdf("polishing-subplot-2.pdf", width = 4, height = 4)
b
print(csmall, vp = subvp)
dev.off()
```

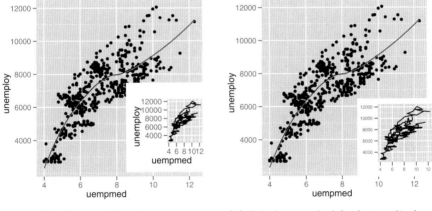

(a) Figure with subplot.   (b) Subplot tweaked for better display.

Fig. 8.9: Two examples of a figure with subplot. It will usually be necessary to tweak the theme settings of the subplot for optimum display.

Note we need to use pdf() (or png() etc.) to save the plots to disk because ggsave() only saves a single plot.

8.4.2 Rectangular grids

A more complicated scenario is when you want to arrange a number of plots in a rectangular grid. Of course you could create a series of viewports and use what you've learned above, but doing all the calculations by hand is cumbersome. A better approach is to use grid.layout(), which sets up a regular grid of viewports with arbitrary heights and widths. You still need to create each viewport, but instead of explicitly specifying the position and size, you can specify the row and column of the layout.

The following example shows how this work. We first create the layout, here a 2×2 grid, then assign it to a viewport and push that viewport on to the plotting device. Now we are ready to draw each plot into its own position on the grid. We create a small function to save some typing, and then draw each plot in the desired place on the grid. You can supply a vector of rows or columns to span a plot over multiple cells. The results are shown in Figure 8.10.

```
pdf("polishing-layout.pdf", width = 8, height = 6)
grid.newpage()
pushViewport(viewport(layout = grid.layout(2, 2)))

vplayout <- function(x, y)
  viewport(layout.pos.row = x, layout.pos.col = y)
print(a, vp = vplayout(1, 1:2))
print(b, vp = vplayout(2, 1))
print(c, vp = vplayout(2, 2))
dev.off()
```

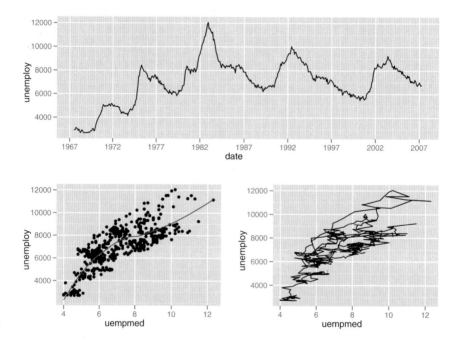

Fig. 8.10: Three plots laid out in a grid using grid.layout().

By default `grid.layout()` makes each cell the same size, but you can use the `widths` and `heights` arguments to make them different sizes. See the documentation for `grid.layout()` for more examples.

Chapter 9

Manipulating data

So far this book has assumed you have your data in a nicely structured data frame ready to feed to `ggplot()` or `qplot()`. If this is not the case, then you'll need to do some transformation.

In Section 9.1, you will learn how to use the `plyr` package to reproduce the statistical transformations performed by the layers, and then in Section 9.2 you will learn a little about "molten" (or long) data, which is useful for time series and parallel coordinates plots, among others. Section 9.3 shows you how to write methods that let you plot objects other than data frames, and demonstrates how `ggplot2` can be used to re-create a more flexible version of the built in linear-model diagnostics.

Data cleaning, manipulation and transformation is a big topic and this chapter only scratches the surface of topics closely related to `ggplot2`. I recommend the following references which go into considerably more depth on this topic:

- *Data Manipulation with R*, by Phil Spector. Published by Springer, 2008.
- "plyr: divide and conquer for data analysis", Hadley Wickham. Available from `http://had.co.nz/plyr`. This is a full description of the package used in Section 9.1.
- "Reshaping data with the reshape package", Hadley Wickham. *Journal of Statistical Software*, 21(12), 2007. `http://www.jstatsoft.org/v21/i12/`. This describes the complement of the melt function used in Section 9.2, which can be used like pivot tables to create a wide range of data summaries and rearrangements.

9.1 An introduction to plyr

With faceting, `ggplot2` makes it very easy to create identical plots for different subsets of your data. This section introduces `ddply()` from the `plyr` package, a function that makes it easy to do the same thing for numerical summaries. `plyr` provides a comprehensive suite of tools for breaking up complicated

H. Wickham, *ggplot2*, Use R, DOI 10.1007/978-0-387-98141-3_9,

data structures into pieces, processing each piece and then joining the results back together. The `plyr` package as a whole provides tools for breaking and combining lists, arrays and data frames. Here we will focus on the `ddply()` function which breaks up a data frame into subsets based on row values, applies a function to each subset and then joins the results back into a data frame. The basic syntax is `ddply(.data, .variables, .fun, ...)`, where

- `.data` is the dataset to break up (e.g., the data that you are plotting).
- `.variables` is a description of the grouping variables used to break up the dataset. This is written like `.(var1, var2)`, and to match the plot should contain all the grouping and faceting variables that you've used in the plot.
- `.fun` is the summary function you want to use. The function can return a vector or data frame. The result does not need to contain the grouping variables: these will be added on automatically if they're needed. The result can be a much reduced aggregated dataset (maybe even one number), or the original data modified or expanded in some way.

More information and examples are available in the documentation, `?ddply`, and on the package website, `http://had.co.nz/plyr`. The following examples show a few useful summary functions that solve common data manipulation problems.

- Using `subset()` allows you to select the top (or bottom) n (or x%) of observations in each group, or observations above (or below) some group-specific threshold:

```
# Select the smallest diamond in each colour
ddply(diamonds, .(color), subset, carat == min(carat))

# Select the two smallest diamonds
ddply(diamonds, .(color), subset, order(carat) <= 2)

# Select the 1% largest diamonds in each group
ddply(diamonds, .(color), subset, carat >
  quantile(carat, 0.99))

# Select all diamonds bigger than the group average
ddply(diamonds, .(color), subset, price > mean(price))
```

- Using `transform()` allows you to perform group-wise transformations with very little work. This is particularly useful if you want to add new variables that are calculated on a per-group level, such as a per-group standardisation. Section 9.2.1 shows another use of this technique for standardising time series to a common scale.

```
# Within each colour, scale price to mean 0 and variance 1
ddply(diamonds, .(color), transform, price = scale(price))
```

```
# Subtract off group mean
ddply(diamonds, .(color), transform,
  price = price - mean(price))
```

- If you want to apply a function to every column in the data frame, you might find the `colwise()` function handy. This function converts a function that operates on vectors to a function that operates column-wise on data frames. This is rather different than most functions: instead of returning a vector of numbers, `colwise()` returns a new function. The following example creates a function to count the number of missing values in a vector and then shows how we can use `colwise()` to apply it to every column in a data frame.

```
> nmissing <- function(x) sum(is.na(x))
> nmissing(msleep$name)
[1] 0
> nmissing(msleep$brainwt)
[1] 27
>
> nmissing_df <- colwise(nmissing)
> nmissing_df(msleep)
  name genus vore order conservation sleep_total sleep_rem
1    0     0    7     0           29           0        22
  sleep_cycle awake brainwt bodywt
1          51     0      27      0
> # This is shorthand for the previous two steps
> colwise(nmissing)(msleep)
  name genus vore order conservation sleep_total sleep_rem
1    0     0    7     0           29           0        22
  sleep_cycle awake brainwt bodywt
1          51     0      27      0
```

The specialised version `numcolwise()` does the same thing, but works only with numeric columns. For example, `numcolwise(median)` will calculate a median for every numeric column, or `numcolwise(quantile)` will calculate quantiles for every numeric column. Similarly, `catcolwise()` only works with categorical columns.

```
> msleep2 <- msleep[, -6] # Remove a column to save space
> numcolwise(median)(msleep2, na.rm = T)
  sleep_rem sleep_cycle awake brainwt bodywt
1       1.5        0.33    14   0.012    1.7
> numcolwise(quantile)(msleep2, na.rm = T)
     sleep_rem sleep_cycle awake brainwt  bodywt
0%         0.1        0.12   4.1 0.00014 5.0e-03
```

```
25%           0.9          0.18   10.2 0.00290 1.7e-01
50%           1.5          0.33   13.9 0.01240 1.7e+00
75%           2.4          0.58   16.1 0.12550 4.2e+01
100%          6.6          1.50   22.1 5.71200 6.7e+03
> numcolwise(quantile)(msleep2, probs = c(0.25, 0.75),
+   na.rm = T)
     sleep_rem sleep_cycle awake brainwt bodywt
25%        0.9          0.18    10  0.0029   0.17
75%        2.4          0.58    16  0.1255  41.75
```

Combined with ddply, this makes it easy to produce per-group summaries:

```
> ddply(msleep2, .(vore), numcolwise(median), na.rm = T)
     vore sleep_rem sleep_cycle awake brainwt bodywt
1   carni      1.95        0.38  13.6  0.0445 20.490
2   herbi      0.95        0.22  13.7  0.0123  1.225
3 insecti      3.00        0.17   5.9  0.0012  0.075
4    omni      1.85        0.50  14.1  0.0066  0.950
5    <NA>      2.00        0.18  13.4  0.0030  0.122
> ddply(msleep2, .(vore), numcolwise(mean), na.rm = T)
     vore sleep_rem sleep_cycle awake brainwt bodywt
1   carni       2.3        0.37    14  0.0793  90.75
2   herbi       1.4        0.42    14  0.6216 366.88
3 insecti       3.5        0.16     9  0.0215  12.92
4    omni       2.0        0.59    13  0.1457  12.72
5    <NA>       1.9        0.18    14  0.0076   0.86
```

- If none of the previous shortcuts is appropriate, make your own summary function which takes a data frame as input and returns an appropriately summarised data frame as output. The following function calculates the rank correlation of price and carat and compares it to the regular correlation of the logged values.

```
> my_summary <- function(df) {
+   with(df, data.frame(
+     pc_cor = cor(price, carat, method = "spearman"),
+     lpc_cor = cor(log(price), log(carat))
+   ))
+ }
> ddply(diamonds, .(cut), my_summary)
        cut pc_cor lpc_cor
1      Fair   0.91    0.91
2      Good   0.96    0.97
3 Very Good   0.97    0.97
4   Premium   0.96    0.97
```

```
5     Ideal   0.95    0.97
> ddply(diamonds, .(color), my_summary)
  color pc_cor lpc_cor
1    D   0.96    0.96
2    E   0.96    0.96
3    F   0.96    0.96
4    G   0.96    0.97
5    H   0.97    0.98
6    I   0.98    0.99
7    J   0.98    0.99
```

Note how our summary function did not need to output the group variables. This makes it much easier to aggregate over different groups.

The common pattern of all these problems is that they are easy to solve if we have the right subset. Often the solution for a single case might be a single line of code. The difficulty comes when we want to apply the function to multiple subsets and then correctly join back up the results. This may take a lot of code, especially if you want to preserve group labels. ddply() takes care of all this for you.

The following case study shows how you can use plyr to reproduce the statistical summaries produced by ggplot2. This is useful if you want to save them to disk or apply them to other datasets. It's also useful to be able to check that ggplot2 is doing exactly what you think!

9.1.1 Fitting multiple models

In this section, we'll work through the process of generating the smoothed data produced by stat_smooth. This process will be the same for any other statistic, and should allow you to produce more complex summaries that ggplot2 can't produce by itself. Figure 9.1 shows the group-wise smoothes produced by the following code.

```
qplot(carat, price, data = diamonds, geom = "smooth",
  colour = color)
dense <- subset(diamonds, carat < 2)
qplot(carat, price, data = dense, geom = "smooth",
  colour = color,  fullrange = TRUE)
```

How can we re-create this by hand? First we read the stat_smooth() documentation to determine what the model is: for large data it's gam(y ~ s(x, bs = "cs")). To get the same output as stat_smooth(), we need to fit the model, then predict it on an evenly spaced grid of points. This task is performed by the smooth() function in the following code. Once we have written this function it is straightforward to apply it to each diamond colour using ddply().

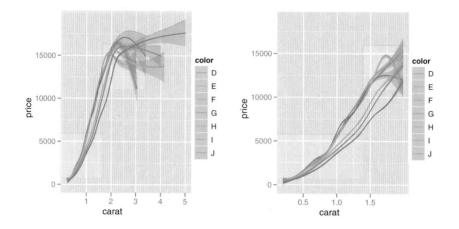

Fig. 9.1: A plot showing the smoothed trends for price vs. carat for each colour of diamonds. With the full range of carats (left), the standard errors balloon after around two carats because there are relatively few diamonds of that size. Restricting attention to diamonds of less than two carats (right) focuses on the region where we have plenty of data.

Figure 9.2 shows the results of this work, which are identical to what we got with ggplot2 doing all the work.

```
library(mgcv)
smooth <- function(df) {
  mod <- gam(price ~ s(carat, bs = "cs"), data = df)
  grid <- data.frame(carat = seq(0.2, 2, length = 50))
  pred <- predict(mod, grid, se = T)

  grid$price <- pred$fit
  grid$se <- pred$se.fit
  grid
}
smoothes <- ddply(dense, .(color), smooth)
qplot(carat, price, data = smoothes, colour = color,
  geom = "line")
qplot(carat, price, data = smoothes, colour = color,
  geom = "smooth", ymax = price + 2 * se, ymin = price - 2 * se)
```

Doing the summary by hand gives you much more flexibility to fit models where the grouping factor is explicitly included as a covariate. For example, the following model models price as a non-linear function of carat, plus a constant term for each colour. It's not a very good model as it predicts negative prices for small, poor-quality diamonds, but it's a starting point for a better model.

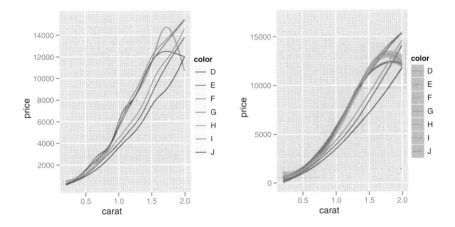

Fig. 9.2: Figure 9.1 with all statistical calculations performed by hand. The predicted values (left), and with standard errors (right).

```
> mod <- gam(price ~ s(carat, bs = "cs") + color, data = dense)
> grid <- with(diamonds, expand.grid(
+   carat = seq(0.2, 2, length = 50),
+   color = levels(color)
+ ))
> grid$pred <- predict(mod, grid)
> qplot(carat, pred, data = grid, colour = color, geom = "line")
```

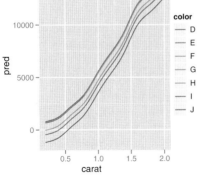

See also Sections 4.9.3 and 5.8 for other ways of combining models and data.

9.2 Converting data from wide to long

In `ggplot2` graphics, groups are defined by rows, not by columns. This makes it easy to draw a line for each group defined by the value of a variable (or set of variables) but difficult to draw a separate line for each variable. In this section you will learn how to transform your data to a form in which you can draw a line for each variable. This transformation converts from "wide" data to "long" data, where each variable now occupies its own set of rows.

To perform this transformation we will use the `melt()` function from the `reshape` package (Wickham, 2007). Reshape also provides the `cast()` function to flexibly reshape and aggregate data, which you may want to read about yourself. Table 9.1 gives an example. The `melt()` function has three arguments:

- `data`: the data frame you want to convert to long form.
- `id.vars`: Identifier (id) variables identify the unit that measurements take place on. Id variables are usually discrete, and are typically fixed by design. In ANOVA notation (Y_{ijk}), id variables are the indices on the variables (i, j, k); in database notation, id variables are a composite primary key.
- `measure.vars`: Measured variables represent what is measured on that unit (Y). These will be the variables that you want to display simultaneously on the plot.

If you're familiar with Wilkinson's grammar of graphics, you might wonder why there is no equivalent to the algebra. There is no equivalent to the algebra within `ggplot2` itself because there are many other facilities for transforming data in R, and it is in line with the `ggplot2` philosophy of keeping data transformation and visualisation as separate as possible.

The following sections explore two important uses of molten data in more detail: plotting multiple time series and creating parallel coordinate plots. You will also see how to use `ddply()` to rescale the variables, and learn about the features of `ggplot2` that are most useful in conjunction with this sort of data.

9.2.1 Multiple time series

Take the `economics` dataset. It contains information about monthly economic data like the number of people unemployed (`unemploy`) and the median length of unemployment (`uempmed`). We might expect these two variables to be related. Each of these variables is stored in a column, which makes it easy to compare them with a scatterplot, and draw individual time series, as shown in Figure 9.3. But what if we want to see the time series simultaneously?

One way is to build up the plot with a different layer for each variable, as you saw in Section 6.4.4. However, this quickly becomes tedious when you have many variables, and a better alternative is to melt the data into a long format and then visualise that. In the molten data the time series have their

date	pce	pop
1967-06-30	508	198,712
1967-07-31	511	198,911
1967-08-31	517	199,113
1967-09-30	513	199,311
1967-10-31	518	199,498
1967-11-30	526	199,657

date	variable	value
1967-06-30	pce	508
1967-07-31	pce	511
1967-08-31	pce	517
1967-09-30	pce	513
1967-10-31	pce	518
1967-11-30	pce	526
1967-06-30	pop	198,712
1967-07-31	pop	198,911
1967-08-31	pop	199,113
1967-09-30	pop	199,311
1967-10-31	pop	199,498
1967-11-30	pop	199,657

Table 9.1: Economics data in wide, left, and long, right, formats. The data stored in each table is equivalent, just the arrangement is different. It it easy to use the wider format with ggplot2 to produce a line for each variable.

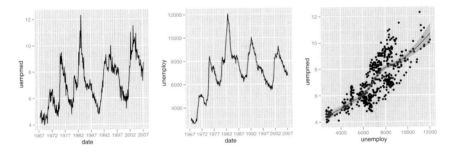

Fig. 9.3: When the economics dataset is stored in wide format, it is easy to create separate time series plots for each variable (left and centre), and easy to create scatterplots comparing them (right).

value stored in the value variable and we can distinguish between them with the variable variable. The code below shows these two alternatives. The plots they produce are very similar, and are shown in Figure 9.4.

```
ggplot(economics, aes(date)) +
  geom_line(aes(y = unemploy, colour = "unemploy")) +
  geom_line(aes(y = uempmed, colour = "uempmed")) +
  scale_colour_hue("variable")

emp <- melt(economics, id = "date",
  measure = c("unemploy", "uempmed"))
qplot(date, value, data = emp, geom = "line", colour = variable)
```

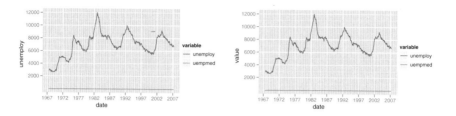

Fig. 9.4: The two methods of displaying both series on a single plot produce identical plots, but using long data is much easier when you have many variables. The series have radically different scales, so we only see the pattern in the unemploy variable. You might not even notice uempmed unless you're paying close attention: it's the line at the bottom of the plot.

There is a problem with these plots: the two variables have radically different scales, and so the series for uempmed appears as a flat line at the bottom of the plot. There is no way to produce a plot with two axes in ggplot2 because this type of plot is fundamentally misleading. Instead there are two perceptually well-founded alternatives: rescale the variables to have a common range, or use faceting with free scales. These alternatives are created with the code below and are shown in Figure 9.5.

```
range01 <- function(x) {
  rng <- range(x, na.rm = TRUE)
  (x - rng[1]) / diff(rng)
}
emp2 <- ddply(emp, .(variable), transform, value = range01(value))
qplot(date, value, data = emp2, geom = "line",
  colour = variable, linetype = variable)
qplot(date, value, data = emp, geom = "line") +
  facet_grid(variable ~ ., scales = "free_y")
```

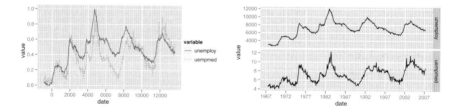

Fig. 9.5: When the series have very different scales we have two alternatives: left, rescale the variables to a common scale, or right, display the variables on separate facets and using free scales.

9.2.2 Parallel coordinates plot

In a similar manner, we can use molten data to create a parallel coordinates plot (Inselberg, 1985; Wegman, 1990), which has the "variable" variable on the x axis and value on the y axis. We need a new variable to record the row that each observation came from, which is used as a grouping variable for the lines (so we get one line per observation). The easiest value to use for this is the data frame rownames, and we give it an unusual name .row, so we don't squash any of the existing variables. Once we have the data in this form, creating a parallel coordinates plot is easy.

The following code does exactly that for the ratings of 840 movies with over 10,000 votes. This dataset has a moderate number of variables (10) and many cases, and will allow us to experiment with a common technique for dealing with large data in parallel coordinates plots: transparency and clustering. Each variable gives the proportion of votes given to each rating between 0 (very bad) and 10 (very good). Since this data is already on a common scale we don't need to rescale it, but in general, we would need to use the technique from the previous section to ensure the variables are comparable. This is particularly important if we are going to use other multidimensional techniques to analyse the data.

```
popular <- subset(movies, votes > 1e4)
ratings <- popular[, 7:16]
ratings$.row <- rownames(ratings)
molten <- melt(ratings, id = ".row")
```

Once the data is in this form, creating a parallel coordinates plot is easy. All we need is a line plot with variable on the x axis, value on the y axis and the lines grouped by .row. This data needs a few tweaks to the default because the values are highly discrete. In the following code, we experiment with jittering and alpha blending to better display where the bulk of the movies lie. The results are shown in Figure 9.6. Most are rated as sevens or eights by around 25% of voters, with a few exceptional movies getting 35% of more perfect 10s. However, the large number of lines makes it difficult to distinguish individual movies and it's hard to draw firm conclusions.

```
pcp <- ggplot(molten, aes(variable, value, group = .row))
pcp + geom_line()
pcp + geom_line(colour = alpha("black", 1 / 20))
jit <- position_jitter(width = 0.25, height = 2.5)
pcp + geom_line(position = jit)
pcp + geom_line(colour = alpha("black", 1 / 20), position = jit)
```

To make the patterns more clear we will cluster the movies into groups of similar rating patterns. The following code uses kmeans clustering (Hartigan and Wong, 1979) to produce six groups of similar movies. To make the clusters

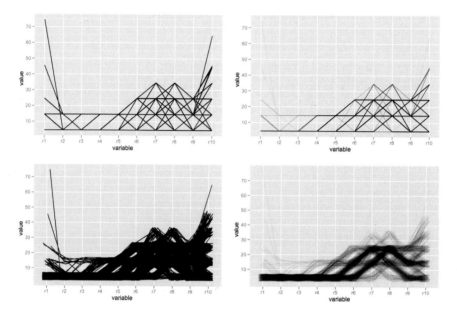

Fig. 9.6: Variants on the parallel coordinates plot to better display the patterns in this highly discrete data. To improve the default pcp (top left) we experiment with alpha blending (top right), jittering (bottom left) and then both together (bottom right).

a little more interpretable, they are relabelled so that cluster 1 has the lowest average rating and cluster six the highest.

```
cl <- kmeans(ratings[1:10], 6)
ratings$cluster <- reorder(factor(cl$cluster), popular$rating)
levels(ratings$cluster) <- seq_along(levels(ratings$cluster))
molten <- melt(ratings, id = c(".row", "cluster"))
```

There are many different ways that we can visualise the result of this clustering. One popular method is shown in Figure 9.7 where line colour is mapped to group membership. This plot is supplemented with a plot that just shows averages for each group. These plots are both straightforward to create, as shown in the following code.

```
pcp_cl <- ggplot(molten,
  aes(variable, value, group = .row, colour = cluster))
pcp_cl + geom_line(position = jit, alpha = 1/5)
pcp_cl + stat_summary(aes(group = cluster), fun.y = mean,
  geom = "line")
```

These plots are good for showing the differences between groups, but they don't tell us a lot about whether we've done a good job clustering the data.

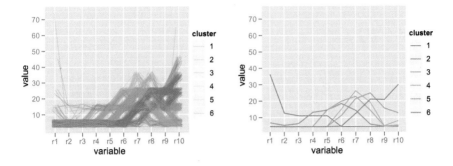

Fig. 9.7: Displaying cluster membership on a parallel coordinates plot. (Left) Individual movies coloured by group membership and (right) group means.

Figure 9.8 uses faceting to display each group in its own panel. This plot highlights the variation within many of the groups, suggesting that perhaps more clusters would be appropriate.

```
pcp_cl + geom_line(position = jit, colour = alpha("black", 1/5)) +
    facet_wrap(~ cluster)
```

9.3 ggplot() methods

ggplot() is a generic function, with different methods for different types of data. The most common input, and what we have used until now, is a data frame. As with base and lattice graphics, it is possible to extend ggplot() to work with other types of data. However, the way this works with ggplot2 is fundamentally different: ggplot2 will not give you a complete plot, but instead will give you the tools you need to make any plot you desire.

This process is mediated by the fortify() method, which takes an object, and optional data frame, and creates a version of the object in a form suitable for plotting with ggplot2, i.e., as a data frame. The name fortify comes from thinking about combining a model with its data: the model fortifies the data, and the data fortifies the model, and the result can be used to simultaneously visualise the model and the data. An example will make this concrete, as you will see when we describe the fortify method for linear models.

This section describes how the fortify() method works, and how you can create new methods that are aligned with the ggplot2 philosophy. The most important philosophical consideration is that data transformation and display should be kept as separate as possible. This maximises reusability, as you are no longer trapped into the single display that the author envisaged.

These different types of input also work with qplot(): remember that qplot() is just a thin wrapper around ggplot().

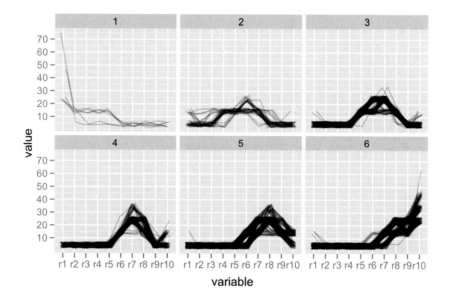

Fig. 9.8: Faceting allows us to display each group in its own panel, highlighting the fact that there seems to be considerable variation within each group, and suggesting that we need more groups in our clustering.

9.3.1 Linear models

Currently, `ggplot2` provides only one fortify method, for linear models. Here we'll show how this method works, and how you can use it to create tailored plots for better understanding your data and models. Figure 9.9 shows the output of `plot.lm()` for a simple model. The graphics are a set of pre-chosen model summary plots. These are useful for particular problems, but are completely inflexible: there is no way to modify them apart from opening up the source code for `plot.lm()` and modifying it. This is hard because the data transformation and display are inextricably entangled, making the code difficult to understand.

The `ggplot2` approach completely separates data transformation and display. The `fortify()` method does the transformation, and then we use `ggplot2` as usual to create the display that we want. Currently `fortify()` adds the variables listed in Table 9.2 to the original dataset. These are basically all the variables that `plot.lm()` creates in order to produce its summary plots. The variables have a leading . (full stop) in their names, so there is little risk that they will clobber variables already in the dataset.

To demonstrate these techniques, we're going to fit the very simple model with code below, which also creates the plot in Figure 9.10. This model clearly doesn't fit the data well, so we should be able to use model diagnostics to

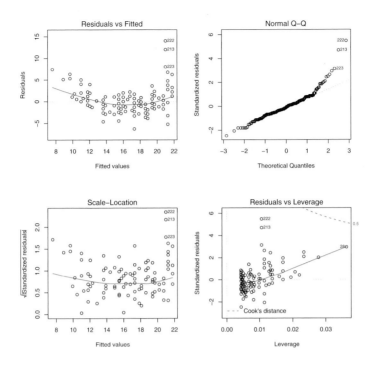

Fig. 9.9: The output from `plot.lm()` for a simple model.

Variable	Description
.cooksd	Cook's distances
.fitted	Fitted values
.hat	Diagonal of the hat matrix
.resid	Residuals
.sigma	Estimate of residual standard deviation when corresponding observation is dropped from model
.stdresid	Standardised residuals

Table 9.2: The diagnostic variables that `fortify.lm()` assembles and adds to the model data.

figure out how to improve it. A sample of the output from fortifying this model is shown in Table 9.3. Because we didn't supply the original data frame, it contains the two variables used in the model as well as the six diagnostic variables. It's easy to see exactly what data our plot will be working with and we could easily add more variables if we wanted.

```
qplot(displ, cty, data = mpg) + geom_smooth(method = "lm")
mpgmod <- lm(cty ~ displ, data = mpg)
```

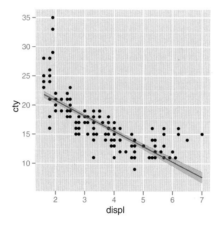

Fig. 9.10: A simple linear model that doesn't fit the data very well.

cty	displ	.hat	.sigma	.cooksd	.fitted	.resid	.stdresid
18	1.80	0.01	2.56	0.01	21.26	-3.26	-1.28
21	1.80	0.01	2.57	0.00	21.26	-0.26	-0.10
20	2.00	0.01	2.57	0.00	20.73	-0.73	-0.29
21	2.00	0.01	2.57	0.00	20.73	0.27	0.11
16	2.80	0.01	2.57	0.00	18.63	-2.63	-1.03
18	2.80	0.01	2.57	0.00	18.63	-0.63	-0.24

Table 9.3: The output of `fortify(mpgmod)` contains the two variables used in the model (`cty` and `displ`), and the six diagnostic variables described above.

With a fortified dataset in hand we can easily re-create the plots produced by `plot.lm()`, and even better, we can adapt them to our needs. The example below shows how we can re-create and then extend the first plot produced by `plot.lm()`. Once we have the basic plot we can easily enhance it: use standardised residuals instead of raw residuals, or make size proportional to Cook's distance. The results are shown in Figure 9.11.

```
mod <- lm(cty ~ displ, data = mpg)
basic <- ggplot(mod, aes(.fitted, .resid)) +
  geom_hline(yintercept = 0, colour = "grey50", size = 0.5) +
  geom_point() +
  geom_smooth(size = 0.5, se = F)
basic
basic + aes(y = .stdresid)
basic + aes(size = .cooksd) + scale_area("Cook's distance")
```

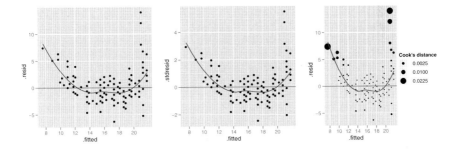

Fig. 9.11: (Left) Basic fitted values-residual plot. (Middle) With standardised residuals. (Right) With size proportional to Cook's distance. It is easy to modify the basic plots when we have access to all of the data.

Additionally, we can fortify the whole dataset and add to the plot variables that are in the original data but not in the model. This helps us to understand what variables are useful to improve the model. Figure 9.12 colours the residuals by the number of cylinders, and suggests that this variable would be good to add to the model: within each cylinder group, the pattern is close to linear.

```
full <- basic %+% fortify(mod, mpg)
full + aes(colour = factor(cyl))
full + aes(displ, colour = factor(cyl))
```

9.3.2 Writing your own

To write your own fortify() method, you will need to think about what variables are most useful for model diagnosis, and how they should be returned to the user. The method for linear models adds them on to the original data frame, but this might not be the best approach in other circumstances, and

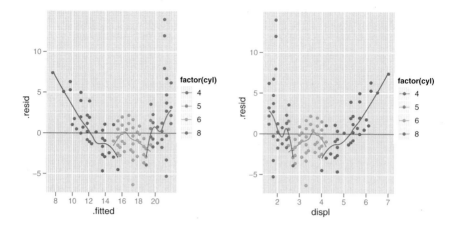

Fig. 9.12: Adding variables from the original data can be enlightening. Here when we add the number of cylinders we see that instead of a curvi-linear relationship between displacement and city mpg, it is essentially linear, conditional on the number of cylinders.

you may instead want to return a list of data frames giving information at different levels of aggregation.

You can also use `fortify()` with non-model functions. The following example shows how we could write a `fortify()` method to make it easier to add images to your plots. The **EBImage** from bioconductor is used to get the image into R, and then the fortify method converts it into a form (a data frame) that **ggplot2** can render. Should you even need a picture of me on your plot, the following code will allow you to do so.

```
fortify.Image <- function(model, data, ...) {
  colours <- channel(model, "x11")[,,]
  colours <- colours[, rev(seq_len(ncol(colours)))]
  melt(colours, c("x", "y"))
}

library(EBImage)
img <- readImage("http://had.co.nz/me.jpg", TrueColor)
qplot(x, y, data = img, fill = value, geom="tile") +
  scale_fill_identity() + coord_equal()
```

This approach cleanly separates the display of the data from its production, and dramatically improves reuse. However, it does not provide any conveniently pre-packaged functions. If you want to create a diagnostic plot for a linear model you have to assemble all the pieces yourself. Once you have the basic structure in place, so that people can always dig back down and alter the individual pieces, you can write a function that joins all the components together in a useful way. See Section 10.4 for some pointers on how to do this.

Chapter 10

Reducing duplication

10.1 Introduction

A major requirement of a good data analysis is flexibility. If the data changes, or you discover something that makes you rethink your basic assumptions, you need to be able to easily change many plots at once. The main inhibitor of flexibility is duplication. If you have the same plotting statement repeated over and over again, you have to make the same change in many different places. Often just the thought of making all those changes is exhausting!

This chapter describes three ways of reducing duplication. In Section 10.2, you will learn how to iteratively modify the previous plot, allowing you to build on top of your previous work without having to retype a lot of code. Section 10.3 will show you how to produce plot "templates" that encapsulate repeated components that are defined once and used in many different places. Finally, 10.4 talks about how to create functions that create or modify plots.

10.2 Iteration

Whenever you create or modify a plot, ggplot2 saves a copy of the result so you can refer to it in later expressions. You can access this plot with last_plot(). This is useful in interactive work as you can start with a basic plot and then iteratively add layers and tweak the scales until you get to the final result. The following code demonstrates iteratively zooming in on a plot to find a region of interest, and then adding a layer which highlights something interesting that we have found: very few diamonds have equal x and y dimensions. The plots are shown in Figure 10.1.

```
qplot(x, y, data = diamonds, na.rm = TRUE)
last_plot() + xlim(3, 11) + ylim(3, 11)
last_plot() + xlim(4, 10) + ylim(4, 10)
last_plot() + xlim(4, 5) + ylim(4, 5)
last_plot() + xlim(4, 4.5) + ylim(4, 4.5)
```

H. Wickham, *ggplot2*, Use R, DOI 10.1007/978-0-387-98141-3_10,
© Springer Science+Business Media, LLC 2009

```
last_plot() + geom_abline(colour = "red")
```

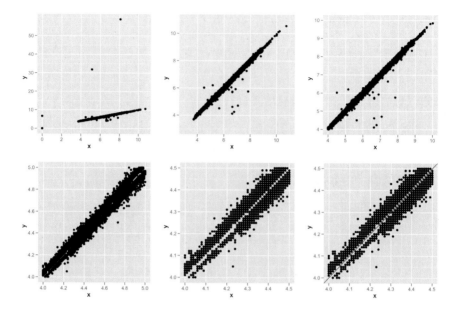

Fig. 10.1: When "zooming" in on the plot, it's useful to use `last_plot()` iteratively to quickly find the best view. The final plot adds a line with slope 1 and intercept 0, confirming it is the square diamonds that are missing.

Once you have tweaked the plot to your liking, it's a good idea to go back and create a single expression that generates your final plot. This is important as when you come back to the plot, you'll be able to re-create the plot quickly, without having to step through your original process. You many want to add a comment to your code to indicate exactly why you chose that final plot. This is good practice in general for R code: after experimenting interactively, you always want to create a source file that re-creates your analysis. The following code shows the final plot after our interactive modifications above.

```
qplot(x, y, data = diamonds, na.rm = T) +
  geom_abline(colour = "red") +
  xlim(4, 4.5) + ylim(4, 4.5)
```

10.3 Plot templates

Each component of a `ggplot2` plot is its own object and can be created, stored and applied independently to a plot. This makes it possible to create reusable components that can automate common tasks and helps to offset the cost of typing the long function names. The following example creates some colour scales and then applies them to plots. The results are shown in Figure 10.2.

```
gradient_rb <- scale_colour_gradient(low = "red", high = "blue")
qplot(cty, hwy, data = mpg, colour = displ) + gradient_rb
qplot(bodywt, brainwt, data = msleep, colour = awake, log="xy") +
  gradient_rb
```

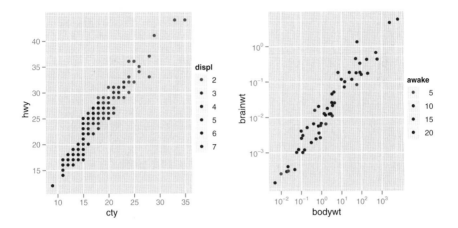

Fig. 10.2: Saving a scale to a variable makes it easy to apply exactly the same scale to multiple plots. You can do the same thing with layers and facets too.

As well as saving single objects, you can also save vectors of **ggplot2** components. Adding a vector of components to a plot is equivalent to adding each component of the vector in turn. The following example creates two continuous scales that can be used to turn off the display of axis labels and ticks. You only need to create these objects once and you can apply them to many different plots, as shown in the code below and Figure 10.3.

```
xquiet <- scale_x_continuous("", breaks = NA)
yquiet <- scale_y_continuous("", breaks = NA)
quiet <- c(xquiet, yquiet)

qplot(mpg, wt, data = mtcars) + quiet
qplot(displ, cty, data = mpg) + quiet
```

Similarly, it's easy to write simple functions that change the defaults of a layer. For example, if you wanted to create a function that added linear models to a plot, you could create a function like the one below. The results are shown in Figure 10.4.

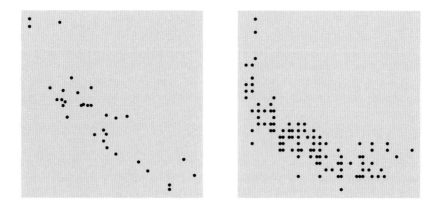

Fig. 10.3: Using "quiet" x and y scales removes the labels and hides ticks and gridlines.

```
geom_lm <- function(formula = y ~ x) {
  geom_smooth(formula = formula, se = FALSE, method = "lm")
}
qplot(mpg, wt, data = mtcars) + geom_lm()
library(splines)
qplot(mpg, wt, data = mtcars) + geom_lm(y ~ ns(x, 3))
```

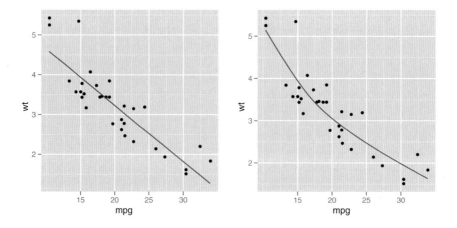

Fig. 10.4: Creating a custom geom function saves typing when creating plots with similar (but not the same) components.

Depending on how complicated your function is, it might even return multiple components in a vector. You can build up arbitrarily complex plots this way, reducing duplication wherever you find it. If you want to create a plot that combines together many different components in a pre-specified way, you might need to write a function that produces the entire plot. This is described in the next section.

10.4 Plot functions

If you are using the same basic plot again and again with different datasets or different parameters, it may be worthwhile to wrap up all the different options into a single function. Maybe you need to perform some data restructuring or transformation, or need to combine the data with a predefined model. In that case you will need to write a function that produces ggplot2 plots. It's hard to give advice on how to go about this because there are so many different possible scenarios, but this section aims to point out some important things to think about.

- Since you're creating the plot within the environment of a function, you need to be extra careful about supplying the data to ggplot() as a data frame, and you need to double check that you haven't accidentally referred to any function local variables in your aesthetic mappings.
- If you want to allow the user to provide their own variables for aesthetic mappings, I'd suggest using aes_string(). This function works just like aes(), but uses strings rather than unevaluated expressions. aes_string("cty", colour = "hwy") is equivalent to aes(cty, colour = hwy). Strings are much easier to work with than expressions.
- As mentioned in Chapter 9, you want to separate your plotting code into a function that does any data transformations and manipulations and a function that creates the plot. Generally, your plotting function should do no data manipulation, just create a plot. The following example shows one way to create parallel coordinate plot function, wrapping up the code used in Section 9.2.2.

```
> pcp_data <- function(df) {
+    numeric <- laply(df, is.numeric)
+    # Rescale numeric columns
+    df[numeric] <- colwise(range01)(df[numeric])
+    # Add row identified
+    df$.row <- rownames(df)
+    # Melt, using non-numeric variables as id vars
+    dfm <- melt(df, id = c(".row", names(df)[!numeric]))
+    # Add pcp to class of the data frame
+    class(dfm) <- c("pcp", class(dfm))
+    dfm
```

```
+ }
> pcp <- function(df, ...) {
+   df <- pcp_data(df)
+   ggplot(df, aes(variable, value)) + geom_line(aes(group =
.row))
+ }
> pcp(mpg)
```

```
> pcp(mpg) + aes(colour = drv)
```

The best example of this technique is `qplot()`, and if you're interesting in writing your own functions I strongly recommend you have a look at the source code for this function and step through it line by line to see how it works. If you've made your way this far through the book you should have a pretty good grasp of all the `ggplot2` related code: most of the complexity is R tricks to correctly interpret all of the possible plot types.

Appendices

Appendix A

Translating between different syntaxes

A.1 Introduction

ggplot2 does not exist in isolation, but is part of a long history of graphical tools in R and elsewhere. This chapter describes how to convert between ggplot2 commands and other plotting systems:

- Within ggplot2, between the qplot() and ggplot() syntaxes, § A.2
- From base graphics, § A.3.
- From lattice graphics, § A.4.
- From GPL, § A.5.

Each section gives a general outline on how to convert between the difference types, followed by a number of examples.

A.2 Translating between qplot and ggplot

Within ggplot2, there are two basic methods to create plots, with qplot() and ggplot(). qplot() is designed primarily for interactive use: it makes a number of assumptions that speed most cases, but when designing multi-layered plots with different data sources it can get in the way. This section describes what those defaults are, and how they map to the fuller ggplot() syntax.

By default, qplot() assumes that you want a scatterplot, i.e., you want to use geom_point().

```
qplot(x, y, data = data)
ggplot(data, aes(x, y)) + geom_point()
```

A.2.1 Aesthetics

If you map additional aesthetics, these will be added to the defaults. With qplot() there is no way to use different aesthetic mappings (or data) in different layers.

H. Wickham, *ggplot2*, Use R, DOI 10.1007/978-0-387-98141-3_BM2,
© Springer Science+Business Media, LLC 2009

```
qplot(x, y, data = data, shape = shape, colour = colour)
ggplot(data, aes(x, y, shape = shape, colour = colour)) +
  geom_point()
```

Aesthetic parameters in `qplot()` always try to map the aesthetic to a variable. If the argument is not a variable but a value, effectively a new column is added to the original dataset with that value. To set an aesthetic to a value and override the default appearance, you surround the value with `I()` in `qplot()`, or pass it as a parameter to the layer. Section 4.5.2 expands on the differences between setting and mapping.

```
qplot(x, y, data = data, colour = I("red"))
ggplot(data, aes(x, y)) + geom_point(colour = "red")
```

A.2.2 Layers

Changing the geom parameter changes the geom added to the plot:

```
qplot(x, y, data = data, geom = "line")
ggplot(data, aes(x, y)) + geom_line()
```

If a vector of multiple geom names is supplied to the geom argument, each geom will be added in turn:

```
qplot(x, y, data = data, geom = c("point", "smooth"))
ggplot(data, aes(x, y)) + geom_point() + geom_smooth()
```

Unlike the rest of **ggplot2**, stats and geoms are independent:

```
qplot(x, y, data = data, stat = "bin")
ggplot(data, aes(x, y)) + geom_point(stat = "bin")
```

Any layer parameters will be passed on to all layers. Most layers will ignore parameters that they don't need.

```
qplot(x, y, data = data, geom = c("point", "smooth"),
  method = "lm")
ggplot(data, aes(x, y)) +
  geom_point(method = "lm") + geom_smooth(method = "lm")
```

A.2.3 Scales and axes

You can control basic properties of the x and y scales with the `xlim`, `ylim`, `xlab` and `ylab` arguments:

```
qplot(x, y, data = data, xlim = c(1, 5), xlab = "my label")
ggplot(data, aes(x, y)) + geom_point() +
  scale_x_continuous("my label", limits = c(1, 5))

qplot(x, y, data = data, xlim = c(1, 5), ylim = c(10, 20))
ggplot(data, aes(x, y)) + geom_point() +
  scale_x_continuous(limits = c(1, 5)) +
  scale_y_continuous(limits = c(10, 20))
```

Like plot(), qplot() has a convenient way of log transforming the axes. There are many other possible transformations that are not accessible from within qplot() see Section 6.4.2 for more details.

```
qplot(x, y, data = data, log="xy")
ggplot(data, aes(x, y)) + geom_point() +
  scale_x_log10() + scale_y_log10()
```

A.2.4 Plot options

qplot() recognises the same options as plot does, and converts them to their ggplot2 equivalents. Section 8.1.2 lists all possible plot options and their effects.

```
qplot(x, y, data = data, main="title", asp = 1)
ggplot(data, aes(x, y)) + geom_point() +
  opts(title = "title", aspect.ratio = 1)
```

A.3 Base graphics

There are two types of graphics functions in base graphics, those that draw complete graphics and those that add to existing graphics.

A.3.1 High-level plotting commands

qplot() has been designed to mimic plot(), and can do the job of all other high-level plotting commands. There are only two graph types from base graphics that cannot be replicated with ggplot2: filled.contour() and persp()

```
plot(x, y);  dotchart(x, y); stripchart(x, y)
qplot(x, y)

plot(x, y, type = "l")
qplot(x, y, geom = "line")
```

```
plot(x, y, type = "s")
qplot(x, y, geom = "step")

plot(x, y, type = "b")
qplot(x, y, geom = c("point", "line"))

boxplot(x, y)
qplot(x, y, geom = "boxplot")

hist(x)
qplot(x, geom = "histogram")

cdplot(x, y)
qplot(x, fill = y, geom = "density", position = "fill")

coplot(y ~ x | a + b)
qplot(x, y, facets = a ~ b)
```

Many of the geoms are parameterised differently than base graphics. For example, `hist()` is parameterised in terms of the number of bins, while `geom_histogram()` is parameterised in terms of the width of each bin.

```
hist(x, bins = 100)
qplot(x, geom = "histogram", binwidth = 1)
```

`qplot()` often requires data in a slightly different format to the base graphics functions. For example, the bar geom works with untabulated data, not tabulated data like `barplot()`; the tile and contour geoms expect data in a data frame, not a matrix like `image()` and `contour()`.

```
barplot(table(x))
qplot(x, geom = "bar")

barplot(x)
qplot(names(x), x, geom = "bar", stat = "identity")

image(x)
qplot(X1, X2, data = melt(x), geom = "tile", fill = value)

contour(x)
qplot(X1, X2, data = melt(x), geom = "contour", fill = value)
```

Generally, the base graphics functions work with individual vectors, not data frames like ggplot2. `qplot()` will try to construct a data frame if one is not specified, but it is not always possible. If you get strange errors, you may need to create the data frame yourself.

```
with(df, plot(x, y))
qplot(x, y, data = df)
```

By default, `qplot()` maps values to aesthetics with a scale. To override this behaviour and set aesthetics, overriding the defaults, you need to use `I()`.

```
plot(x, y, col = "red", cex = 1)
qplot(x, y, colour = I("red"), size = I(1))
```

A.3.2 Low-level drawing

The low-level drawing functions which add to an existing plot are equivalent to adding a new layer in `ggplot2`, described in Table A.1.

Base function	ggplot2 layer
`curve()`	`geom_curve()`
`hline()`	`geom_hline()`
`lines()`	`geom_line()`
`points()`	`geom_point()`
`polygon()`	`geom_polygon()`
`rect()`	`geom_rect()`
`rug()`	`geom_rug()`
`segments()`	`geom_segment()`
`text()`	`geom_text()`
`vline()`	`geom_vline()`
`abline(lm(y ~ x))`	`geom_smooth(method = "lm")`
`lines(density(x))`	`geom_density()`
`lines(loess(x, y))`	`geom_smooth()`

Table A.1: Equivalence between base graphics methods that add on to an existing plot, and layers in `ggplot2`.

```
plot(x, y)
lines(x, y)

qplot(x, y) + geom_line()

# Or, building up piece-meal
qplot(x, y)
last_plot() + geom_line()
```

A.3.3 Legends, axes and grid lines

In `ggplot2`, the appearance of legends and axes is controlled by the scales. Axes are produced by the x and y scales, while all other scales produce legends. See plot themes, Section 8.1, to change the appearance of axes and legends, and, scales, Section 6.5, to change their contents. The appearance of grid lines is controlled by the `grid.major` and `grid.minor` theme options, and their position by the breaks of the x and y scales.

A.3.4 Colour palettes

Instead of global colour palettes, `ggplot2` has scales for individual plots. Much of the time you can rely on the default colour scale (which has somewhat better perceptual properties), but if you want to reuse an existing colour palette, you can use `scale_colour_manual()`. You will need to make sure that the colour is a factor for this to work.

```
palette(rainbow(5))
plot(1:5, 1:5, col = 1:5, pch = 19, cex = 4)

qplot(1:5, 1:5, col = factor(1:5), size = I(4))
last_plot() + scale_colour_manual(values = rainbow(5))
```

In `ggplot2`, you can also use palettes with continuous values, with intermediate values being linearly interpolated.

```
qplot(0:100, 0:100, col = 0:100, size = I(4)) +
  scale_colour_gradientn(colours = rainbow(7))
last_plot() +
  scale_colour_gradientn(colours = terrain.colors(7))
```

A.3.5 Graphical parameters

The majority of `par` settings have some analogue within the theme system, or in the defaults of the geoms and scales. The appearance plot border drawn by `box()` can be controlled in a similar way by the `panel.background` and `plot.background` theme elements. Instead of using `title()`, the plot title is set with the `title` option.

A.4 Lattice graphics

The major difference between lattice and `ggplot2` is that lattice uses a formula-based interface. `ggplot2` does not because the formula does not generalise well to more complicated situations.

```
xyplot(rating ~ year, data=movies)
qplot(year, rating, data=movies)

xyplot(rating ~ year | Comedy + Action, data = movies)
qplot(year, rating, data = movies, facets = ~ Comedy + Action)
# Or maybe
qplot(year, rating, data = movies, facets = Comedy ~ Action)
```

While lattice has many different functions to produce different types of graphics (which are all basically equivalent to setting the panel argument), ggplot2 has qplot().

```
stripplot(~ rating, data = movies, jitter.data = TRUE)
qplot(rating, 1, data = movies, geom = "jitter")

histogram(~ rating, data = movies)
qplot(rating, data = movies, geom = "histogram")

bwplot(Comedy ~ rating ,data = movies)
qplot(factor(Comedy), rating, data = movies, type = "boxplot")

xyplot(wt ~ mpg, mtcars, type = c("p","smooth"))
qplot(mpg, wt, data = mtcars, geom = c("point","smooth"))

xyplot(wt ~ mpg, mtcars, type = c("p","r"))
qplot(mpg, wt, data = mtcars, geom = c("point","smooth"),
  method = "lm")
```

The capabilities for scale manipulations are similar in both ggplot2 and lattice, although the syntax is a little different.

```
xyplot(wt ~ mpg | cyl, mtcars, scales = list(y = list(relation
                                     = "free")))
qplot(mpg, wt, data = mtcars) + facet_wrap(~ cyl, scales = "free")

xyplot(wt ~ mpg | cyl, mtcars, scales = list(log = 10))
qplot(mpg, wt, data = mtcars, log = "xy")

xyplot(wt ~ mpg | cyl, mtcars, scales = list(log = 2))
qplot(mpg, wt, data = mtcars) +
  scale_x_log2() + scale_y_log2()

xyplot(wt ~ mpg, mtcars, group = cyl, auto.key = TRUE)
# Map directly to an aesthetic like colour, size, or shape.
qplot(mpg, wt, data = mtcars, colour = cyl)
```

```
xyplot(wt ~ mpg, mtcars, xlim = c(20,30))
# Works like lattice, except you can't specify a different limit
# for each panel/facet
qplot(mpg, wt, data = mtcars, xlim = c(20,30))
```

Both lattice and `ggplot2` have similar options for controlling labels on the plot.

```
xyplot(wt ~ mpg, mtcars,
  xlab = "Miles per gallon", ylab = "Weight",
  main = "Weight-efficiency tradeoff")
qplot(mpg, wt, data = mtcars,
  xlab = "Miles per gallon", ylab = "Weight",
  main = "Weight-efficiency tradeoff")

xyplot(wt ~ mpg, mtcars, aspect = 1)
qplot(mpg, wt, data = mtcars, asp = 1)
```

`par.settings()` is equivalent to + `opts()` and `trellis.options.set()` and `trellis.par.get()` to `theme_set()` and `theme_get()`.

More complicated lattice formulas are equivalent to rearranging the data before using `ggplot2`.

A.5 GPL

The Grammar of Graphics uses two specifications. A concise format is used to caption figures, and a more detailed xml format stored on disk. The following example of the concise format is adapted from Wilkinson (2005, Figure 1.5, page 13).

```
DATA: source("demographics")
DATA: longitude, latitude = map(source("World"))
TRANS: bd = max(birth - death, 0)
COORD: project.mercator()
ELEMENT: point(position(lon * lat), size(bd), color(color.red))
ELEMENT: polygon(position(longitude * latitude))
```

This is relatively simple to adapt to the syntax of `ggplot2`:

- `ggplot()` is used to specify the default data and default aesthetic mappings.
- Data is provided as standard R data.frames existing in the global environment; it does not need to be explicitly loaded. We also use a slightly different world dataset, with columns lat and long. This lets us use the same aesthetic mappings for both datasets. Layers can override the default data and aesthetic mappings provided by the plot.
- We replace TRANS with an explicit transformation by R code.

- ELEMENTs are replaced with layers, which explicitly specify the data source. Each geom has a default statistic which is used to transform the data prior to plotting. For the geoms in this example, the default statistic is the identity function. Fixed aesthetics (the colour red in this example) are supplied as additional arguments to the layer, rather than as special constants.
- The SCALE component has been omitted from this example (so that the defaults are used). In both the ggplot2 and GoG examples, scales are defined by default. In ggplot you can override the defaults by adding a scale object, e.g., scale_colour or scale_size.
- COORD uses a slightly different format. In general, most of the components specifications in ggplot are slightly different to those in GoG, in order to be more familiar to R users.
- Each component is added together with + to create the final plot.

All up the equivalent ggplot2 code is:

```
demographics <- transform(demographics,
  bd = pmax(birth - death, 0))

ggplot(demographic, aes(lon, lat)) +
  geom_polyogon(data = world) +
  geom_point(aes(size = bd), colour = "red") +
  coord_map(projection = "mercator")
```

Appendix B

Aesthetic specifications

This appendix summarises the various formats that `grid` drawing functions take. Most of this information is available scattered throughout the R documentation. This appendix brings it all together in one place.

B.1 Colour

Colours can be specified with:

- A **name**, e.g., `"red"`. The colours are displayed in Figure B.1(a), and can be listed in more detail with `colours()`. The Stowers Institute provides a nice printable pdf that lists all colours: `http://research.stowers-institute.org/efg/R/Color/Chart/`.
- An **rgb specification**, with a string of the form `"#RRGGBB"` where each of the pairs RR, GG, BB consists of two hexadecimal digits giving a value in the range 00 to FF. Partially transparent can be made with `alpha()`, e.g., `alpha("red", 0.5)`.
- An **NA**, for a completely transparent colour.

The functions `rgb()`, `hsv()`, `hcl()` can be used to create colours specified in different colour spaces.

B.2 Line type

Line types can be specified with:

- An **integer** or **name**: 0=blank, 1=solid, 2=dashed, 3=dotted, 4=dotdash, 5=longdash, 6=twodash), illustrated in Figure B.1(b).
- The lengths of on/off stretches of line. This is done with a string of an even number (up to eight) of hexadecimal digits which give the lengths in consecutive positions in the string. For example, the string `"33"` specifies

three units on followed by three off and "3313" specifies three units on followed by three off followed by one on and finally three off.

The five standard dash-dot line types described above correspond to 44, 13, 134, 73 and 2262.

Note that NA is not a valid value for lty.

B.3 Shape

Shapes take four types of values:

- An **integer** in [0, 25], illustrated in Figure B.1(c).
- A **single character**, to use that character as a plotting symbol.
- A **.** to draw the smallest rectangle that is visible (i.e., about one pixel).
- An **NA**, to draw nothing.

While all symbols have a foreground colour, symbols 19–25 also take a background colour (fill).

B.4 Size

Throughout ggplot2, for text height, point size and line width, size is specified in millimetres.

B.5 Justification

Justification of a string (or legend) defines the location within the string that is placed at the given position. There are two values for horizontal and vertical justification. The values can be:

- A **string**: "left", "right", "centre", "center", "bottom", and "top".
- A **number** between 0 and 1, giving the position within the string (from bottom-left corner). These values are demonstrated in Figure B.1(d).

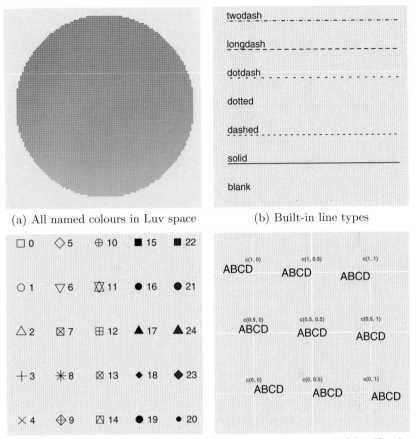

(a) All named colours in Luv space

(b) Built-in line types

(c) R plotting symbols. Colour is black, and fill is blue. Symbol 25 (not shown) is symbol 24 rotated 180 degrees.

(d) Horizontal and vertical justification settings.

Fig. B.1: Examples illustrating different aesthetic settings.

Appendix C

Manipulating plot rendering with `grid`

C.1 Introduction

Sometimes you may need to go beyond the theming system and directly modify the underlying grid graphics output. To do this, you will need a good understanding of grid, as described in "R Graphics" (Murrell, 2005). If you can't get the book, at least read Chapter 5, "The grid graphics model", which is available online for free at `http://www.stat.auckland.ac.nz/~paul/RGraphics/chapter5.pdf`. This appendix outlines the more important viewports and grobs used by `ggplot2` and should be helpful if you need to interact with the grobs produced by `ggplot2`.

C.2 Plot viewports

Viewports define the basic regions of the plot. The structure will vary slightly from plot to plot, depending on the type of faceting used, but the basics will remain the same.

The `panels` viewport contains the meat of the plot: strip labels, axes and faceted panels. The viewports are named according to both their job and their position on the plot. A prefix (listed below) describes the contents of the viewport, and is followed by integer x and y position (counting from bottom left) separated by "_". Figure C.1 illustrates this naming scheme for a 2×2 plot.

- `strip_h`: horizontal strip labels
- `strip_v`: vertical strip labels
- `axis_h`: horizontal axes
- `axis_v`: vertical axes
- `panel`: faceting panels

The `panels` viewport is contained inside the `background` viewport which also contains the following viewports:

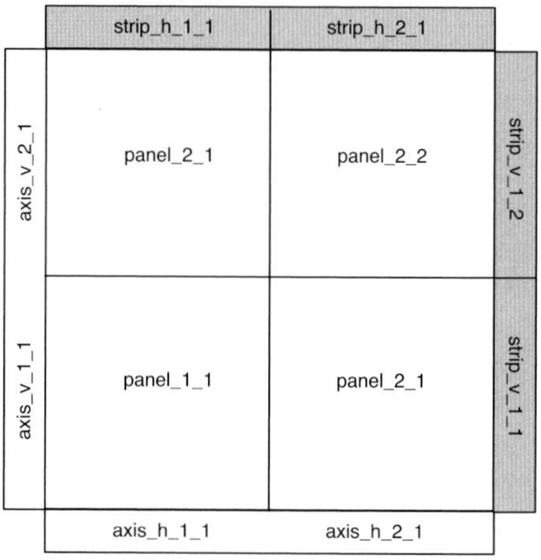

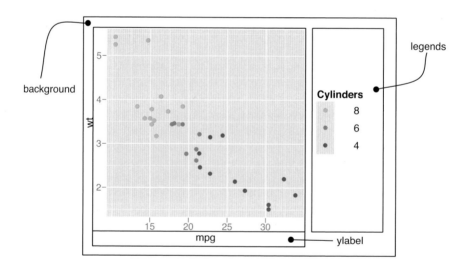

strip_h_1_1
strip_h_2_1
axis_v_2_1
panel_2_1
panel_2_2
strip_v_1_2
axis_v_1_1
panel_1_1
panel_2_1
strip_v_1_1
axis_h_1_1
axis_h_2_1

Fig. C.1: Naming scheming of the panel viewports.

- `title`, `xlabel` and `ylabel`: for the plot title, and x and y axis labels
- `legend_box`: for all of the legends for the plot

Figure C.2 labels a plot with a representative sample of these viewports. To get a list of all viewports on the current plot, run `current.vpTree(all=TRUE)` or `grid.ls(grobs = FALSE, viewports = TRUE)`.

background
legends
Cylinders
8
6
4
wt
5
4
3
2
15
20
25
30
mpg
ylabel

Fig. C.2: Diagram showing the structure and names of viewports.

C.3 Plot grobs

Grob names have three components: the name of the grob, the class of the grob and a unique numeric suffix. The three components are joined together with "." to give a name like `title.text.435` or `ticks.segments.15`. These three components ensure that all grob names are unique, and allow you to select multiple grobs with the same name at the same time. Figure C.3 labels some of these grobs. The grobs are arranged hierarchically, but it's hard to capture this in a diagram. You can see a list of all the grobs in the current plot with `grid.ls()`.

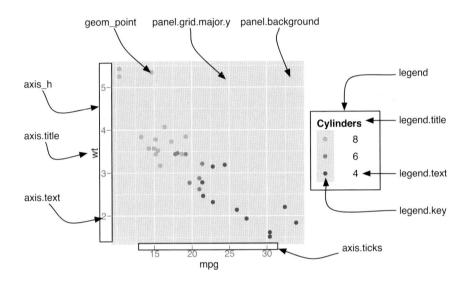

Fig. C.3: A selection of the most important grobs.

C.4 Saving your work

Using `grid.gedit()`, and similar functions, works fine if you are editing the plot on screen, but if you want to save it to disk you need to take some extra steps, or you will end up with multiple pages of output, each showing one change. The key is not to modify the plot on screen, but to modify the plot grob, and then draw it once you have made all the changes.

```
p <- qplot(wt, mpg, data=mtcars, colour=cyl)
# Get the plot grob
grob <- ggplotGrob(p)
# Modify in place
```

```
grob <- geditGrob(grob, gPath("strip","label"), gp=gpar(fontface
                                                    ="bold"))
# Draw it
grid.newpage()
grid.draw(grob)
```

An alternative is make all of the changes on screen, and then use
`dev.copy2pdf()` to copy the final version to disk.

References

A. Azzalini and A. W. Bowman. A look at some data on the Old Faithful geyser. *Applied Statistics*, 39:357–365, 1990.

Cynthia A. Brewer. Color use guidelines for mapping and visualization. In A.M. MacEachren and D.R.F. Taylor, editors, *Visualization in Modern Cartography*, pages 123–147. Elsevier Science, 1994a.

Cynthia A. Brewer. Guidelines for use of the perceptual dimensions of color for mapping and visualization. In *Color Hard Copy and Graphic Arts III, Proceedings of the International Society for Optical Engineering (SPIE), San Jose*, volume 2171, pages 54–63, 1994b.

D. B. Carr, R. J. Littlefield, W. L. Nicholson, and J. S. Littlefield. Scatterplot matrix techniques for large N. *Journal of the American Statistical Association*, 82(398):424–436, 1987.

Dan Carr. Using gray in plots. *ASA Statistical Computing and Graphics Newsletter*, 2(5):11–14, 1994. URL http://www.galaxy.gmu.edu/~dcarr/lib/v5n2.pdf.

Dan Carr. Graphical displays. In Abdel H. El-Shaarawi and Walter W. Piegorsch, editors, *Encyclopedia of Environmetrics*, volume 2, pages 933–960. John Wiley & Sons, 2002. URL http://www.galaxy.gmu.edu/~dcarr/lib/EnvironmentalGraphics.pdf.

Dan Carr and Ru Sun. Using layering and perceptual grouping in statistical graphics. *ASA Statistical Computing and Graphics Newsletter*, 10(1):25–31, 1999.

Dan Carr, Nicholas Lewin-Koh, and Martin Maechler. *hexbin: Hexagonal Binning Routines*, 2008. R package version 1.14.0.

John Chambers, William Cleveland, Beat Kleiner, and Paul Tukey. *Graphical methods for Data Analysis*. Wadsworth, 1983.

William Cleveland. *Visualizing Data*. Hobart Press, 1993a.

William Cleveland. A model for studying display methods of statistical graphics. *Journal of Computational and Graphical Statistics*, 2:323–364, 1993b. URL http://stat.bell-labs.com/doc/93.4.ps.

William Cleveland. *The Elements of Graphing Data*. Hobart Press, 1985.

William S Cleveland and Robert McGill. Graphical perception: The visual decoding of quantitative information on graphical displays of data. *Journal of the Royal Statistical Society. Series A (General)*, 150(3):192–229, 1987.

Dianne Cook and Deborah F. Swayne. *Interactive and Dynamic Graphics for Data Analysis: With Examples Using R and GGobi*. Springer, 2007.

John Fox. *effects: Effect Displays for Linear and Generalized Linear Models*, 2008. URL http://socserv.socsci.mcmaster.ca/jfox/. R package version 1.0-12.

Jr Harrell, Frank E. *Hmisc: Harrell Miscellaneous*, 2008. URL http://biostat.mc.vanderbilt.edu/s/Hmisc. R package version 3.5-2. With contributions from many others.

J. A. Hartigan and M. A. Wong. A k-means clustering algorithm. *Applied Statistics*, 28:100–108, 1979.

A. Inselberg. The plane with parallel coordinates. *The Visual Computer*, 1: 69–91, 1985.

Jim Lemon, Ben Bolker, Sander Oom, Eduardo Klein, Barry Rowlingson, Hadley Wickham, Anupam Tyagi, Olivier Eterradossi, Gabor Grothendieck, Michael Toews, and John Kane. *plotrix: Various plotting functions*, 2008. R package version 2.4-3.

Thomas Lumley. *dichromat: Color schemes for dichromats*, 2007. R package version 1.2-2.

Doug McIlroy. *mapproj: Map Projections*, 2005. R package version 1.1-7.1. Packaged for R by Ray Brownrigg and Thomas P Minka.

David Meyer, Achim Zeileis, and Kurt Hornik. The strucplot framework: Visualizing multi-way contingency tables with vcd. *Journal of Statistical Software*, 17(3):1–48, 2006. URL http://www.jstatsoft.org/v17/i03/.

Paul Murrell. *Investigations in Graphical Statistics*. PhD thesis, The University of Auckland, 1998.

Paul Murrell. *R Graphics*. Chapman & Hall/CRC, 2005.

Naomi Robbins. *Creating More Effective Graphs*. Wiley-Interscience, 2004.

Deepayan Sarkar. *lattice: Lattice Graphics*, 2008a. R package version 0.17-6.

Deepayan Sarkar. *Lattice: Multivariate Data Visualization with R*. Springer, 2008b.

Edward R. Tufte. *Envisioning Information*. Graphics Press, 1990.

Edward R. Tufte. *Visual Explanations*. Graphics Press, 1997.

Edward R. Tufte. *The Visual Display of Quantitative Information*. Graphics Press, 2001.

Edward R. Tufte. *Beautiful Evidence*. Graphics Press, 2006.

John W. Tukey. *Exploratory Data Analysis*. Addison–Wesley, 1977.

Gregory Warnes. *gplots: Various R programming tools for plotting data*, 2007. R package version 2.6.0. Includes R source code and/or documentation contributed by Ben Bolker and Thomas Lumley.

Edward J. Wegman. Hyperdimensional data analysis using parallel coordinates. *Journal of the American Statistical Association*, 85(411):664–675, 1990.

Hadley Wickham. Reshaping data with the reshape package. *Journal of Statistical Software*, 21(12), 2007. URL http://www.jstatsoft.org/v21/i12/paper.

Hadley Wickham. A layered grammar of graphics. *Journal of Computational and Graphical Statistics*, 2009. In press.

Hadley Wickham. *Practical tools for exploring data and models*. PhD thesis, Iowa State University, 2008. URL `http://had.co.nz/thesis`.

Hadley Wickham, Michael Lawrence, Duncan Temple Lang, and Deborah F Swayne. An introduction to rggobi. *R-news*, 8(2):3–7, October 2008. URL `http://CRAN.R-project.org/doc/Rnews/Rnews_2008-2.pdf`.

Leland Wilkinson. *The Grammar of Graphics*. Statistics and Computing. Springer, 2nd edition, 2005.

Achim Zeileis, Kurt Hornik, and Paul Murrell. Escaping RGBland: Selecting colors for statistical graphics. *Computational Statistics & Data Analysis*, 2008. URL `http://statmath.wu-wien.ac.at/~zeileis/papers/Zeileis+Hornik+Murrell-2008.pdf`. Forthcoming.

Index

R code index

A Beginner's Guide to R

Alain F. Zuur, Elena N. Ieno, Erik H.W.G. Meesters, and Den Burg

Based on their extensive experience with teaching R and statistics to applied scientists, the authors provide a beginner's guide to R. To avoid the difficulty of teaching R and statistics at the same time, statistical methods are kept to a minimum. The text covers how to download and install R, import and manage data, elementary plotting, an introduction to functions, advanced plotting, and common beginner mistakes.

2009. Approx. 215 p. Softcover (Use R)
ISBN: 978-0-387-93836-3

R for SAS and SPSS Users

Robert A. Muenchen

Content: Introduction. The five main parts of SAS and SPSS.- Programming conventions.- Typographic conventions.- Installing & updating R.- Running R.- Help and documentation.- Programming language basics.- Data Acquisition.- Selecting Variables - Var, Variables.- Selecting observations - where, if select if, filter.- Selecting both variables and observations.- Converting data structures.- Data management.- Recoding variables. Value labels or formats (& measurement level).- Variable labels.- Generating data. - How R stores data.- Managing your files and workspace.- Graphics overview.- Traditional graphics.- The ggplot2 package.- Statistics.- Summary.- Conclusion.- Appendix A.- Appendix B.- Appendix C.- Bibliography.

2009. XVII, 470 p. Hardcover (Statistics and Computing)
ISBN: 978-0-387-09417-5

Software for Data Analysis

John M. Chambers

This book guides the reader through programming with R, beginning with simple interactive use and progressing by gradual stages, starting with simple functions. The techniques covered include such modern programming enhancements as classes and methods, namespaces, and interfaces to spreadsheets or data bases, as well as computations for data visualization, numerical methods, and the use of text data.

2008. XIV, 498 p. Hardcover (Statistics and Computing)
ISBN: 978-0-387-75935-7

Easy Ways to Order ▶   Call: Toll-Free 1-800-SPRINGER • E-mail: orders-ny@springer.com • Write: Springer, Dept. S8113, PO Box 2485, Secaucus, NJ 07096-2485 • Visit: Your local scientific bookstore or urge your librarian to order.

Printed in the United States of America